Beyond Great Walls

# Beyond Great Walls

ENVIRONMENT, IDENTITY, AND DEVELOPMENT

ON THE CHINESE GRASSLANDS OF

INNER MONGOLIA

*Dee Mack Williams*

STANFORD UNIVERSITY PRESS

STANFORD, CALIFORNIA

Stanford University Press
Stanford, California

© 2002 by the Board of Trustees of the
Leland Stanford Junior University

Printed in the United States of America
on acid-free, archival-quality paper

Library of Congress Cataloging-in-Publication Data

Williams, Dee Mack.
   Beyond great walls : environment, identity, and development on the
Chinese grasslands of Inner Mongolia / Dee Mack Williams.
       p.   cm.
   Includes bibliographical references and index.
   ISBN 0-8047-4278-2 (alk. paper)
   1. Human ecology—China—Inner Mongolia.   2. Ethnicity—China—
Inner Mongolia.   3. Grassland ecology—China—Inner Mongolia.
4. Pastoral systems—China—Inner Mongolia.   5. Land degredation—
China—Inner Mongolia.   6. Environmental degredation—China—Inner
Mongolia.   7. Inner Mongolia (China)—Environmental conditions.
8. Inner Mongolia (China)—Social conditions.   I. Title.
GF657.C6 .W55   2002
333.74'0951'77—dc21                                          2001049812

Original Printing 2002
Last figure below indicates year of this printing:
11   10   09   08   07   06   05   04   03   02

Typeset at Stanford University Press in 10/13 Sabon

# Acknowledgments

Grants from the National Science Foundation and the Committee for Scholarly Communication with China made possible the field research upon which this book is based. I would also like to recognize and express thanks to Professor James Reardon-Anderson for facilitating my field placement, to Dabagan for special assistance during fieldwork, and to all the members of the Grassland Ecosystem Research Station at Wulanaodu for hosting and promoting my research project, especially the directors, Kou Zhenwu and Jiang Deming. I am equally grateful to the residents of Nasihan township who accepted me into their community and enriched my study with their voluntary cooperation.

Subsequent to fieldwork, I received financial support from the American Council of Learned Societies–Chiang Ching-kuo Foundation and the Pacific Cultural Foundation. Also, many scholars helped me to refine my analysis and writing. I especially want to acknowledge my thanks and respect to Professors Michael Dove, Burton Pasternak, Stevan Harrell, Myron Cohen, William Jankowiak, Uradyn Bulag, and Dennis Sheehy for constructive commentary on various drafts of the manuscript. I also benefited from a brief visit with Professor Caroline Humphrey and the Mongolia and Inner Asian Studies Unit at the University of Cambridge. Karin Breiwitz and Bruce Cornelison at the Center for Teaching and Learning at the University of North Carolina at Chapel Hill helped with the final preparation of maps and illustrations. I also thank Nathan MacBrien for his excellent work as my acquisitions and copy editor at Stanford University Press.

Portions of this book have been previously published. Parts of Chapter 2 first appeared in *Modern China* 23(3) (1997): 328–355. Parts of Chapters 4 and 8 first appeared in the *Journal of Asian Studies* 55(3) (1996): 665–691. Parts of Chapter 6 first appeared in *Human Organiza-*

*tion* 55(3) (1996): 307–313. And parts of Chapter 8 first appeared in *Human Ecology* 25(2) (1997): 297–317. I thank Sage Publications, the Association for Asian Studies, the Society for Applied Anthropology, and Plenum Publishing Corporation, respectively, for their permission to reproduce this material here.

Finally, I must acknowledge admiration for Joanne and Jecoliah, my wife and oldest daughter, who accompanied me to a remote field site at great personal sacrifice and encouraged the completion of this project in countless ways. Bella, my youngest daughter, added further inspiration when she later joined the family. They are all emotional co-authors of this work.

# Contents

# Tables

# Illustrations

# Preface

To what environment do you belong? This serious contemporary question confronts people everywhere, both as individuals and as members of a social group. Yet the answers become more complicated and more pressing as humans tread ever more heavily upon the planet and its natural resources. A rapidly changing world forces each of us to consider what kind of nature we prefer, what kind we will accept, and what kind we will resist. The answers depend upon who we imagine ourselves to be, and what we will allow ourselves to become.

This book is about a community caught at an environmental crossroads. On the surface, it is a study of pastoral resource management in the context of ecological crisis. More broadly, it addresses local experiences of modernization and the ways that marginalized people creatively appropriate alien technologies and subvert colonial scientific practices to service their own ethnic identity and cultural renewal. It analyzes a specific setting in order to explore major issues that concern us all.

Since Chinese officials first authorized decollectivization and privatization on national rangelands in the early 1980s, an unfamiliar and disruptive set of land use practices have emerged on the grasslands of Inner Mongolia. My fieldwork among ethnic Mongol herders in a township at the forefront of an expanding enclosure movement reveals that newly erected wire fences are dramatically reshaping the natural environment and redefining access to community resources. Deceptively simple changes in the structure of local space and ecology exacerbate regional problems of land degradation and intensify economic disparities among herding households, setting off a chain of transformation across both the physical and social landscape.

This book describes and analyzes that transformation process and explains the many reasons why the drama should compel our close attention. It documents in detail some of the obvious and more subtle changes that have occurred, and it shows their practical consequences for the lives of local residents. The narrative journey exposes a rather grim rural setting full of remarkable people and extraordinary circumstances—prolonged natural disaster, violent land disputes, mutilation of livestock, rampant alcoholism, chronic cold stress, a surge in criminal activity, symbolic suicide, forced sterilization, migrant labor, adulteration of wool and cashmere, subversion of scientific research, and myriad manifestations of government corruption and abuse of power.

The data and analysis will show that recent government initiatives and the changes that ensued are intimately related to a national modernization project that directly links local residents to an encroaching capitalist world system. They will also show that the reconstruction of local landscape, and the specific problem of land degradation, cannot be fully understood apart from the social context of economic insecurity and political fear, as well as the cultural context of group identity and environmental symbolism. Ideologically informed perceptions of the land prove to be highly relevant in both shaping and contesting international development agendas, national rangeland policies, and the daily practices of local production.

The problems of land degradation in Inner Mongolia are by no means imaginary—they pose real and serious concerns to many different groups of people—but the conventional perceptions and remedial actions imposed by outsiders to improve that reality generally make the situation worse. By presenting the full range of material and symbolic stakes now in play on the Chinese grasslands, this book ultimately demonstrates that human–land interactions involve social dimensions of widely underestimated complexity that we cannot continue to ignore.

# Abbreviations

The following abbreviations are used throughout the text, Notes, and Bibliography:

BMOF    Bureau of the Ministry of Forestry
CAS     Chinese Academy of Sciences
CASS    Chinese Academy of Social Sciences
CCP     Chinese Communist Party
CSC     China State Council
DDR     Department of Desert Research
ECCIA   Environmental and Cultural Conservation in Inner Asia (project)
GEMS    Grassland Ecosystem of the Mongolian Steppe (project)
IDNDR   International Decade for Natural Disaster Reduction
IFAD    International Fund for Agricultural Development
IMAR    Inner Mongolia Autonomous Region
NCPLDP  Northern Pasture and Livestock Development Project
NRC     National Resource Council
PALD    Policy Alternatives for Livestock Development (project)
SEU     sheep equivalent units
UNEP    United Nations Environment Programme
WCED    World Commission on Environment and Development
WHO     World Health Organization

Beyond Great Walls

# ONE

# A Land and People in the Way

North of the Great Wall of China lies a vast expanse of grassland where steep hills punctuate a rising plateau to form the steppes of the Inner Mongolia Autonomous Region (IMAR). It is a geographically remote, economically impoverished, technologically unsophisticated, and culturally distinct part of the developing world. Although there can be no doubt that Chinese military forces securely control the countryside, it is not so easy to identify the prevailing social forces that actually animate the land and people. Complicated economic interests, political histories, social constraints, and cultural values at the local level actively filter the exercise of national government authority and frustrate the disciplinary influences of global capital. Even today, setting out from Beijing and passing beyond the massive stone enclosure that has become the singular icon of Han civilization, one apprehends a distinctive living space opening ahead.

Along with the obvious changes of climate and terrain, human settlement patterns and agricultural activity, a host of more subtle differences—in diet, dress, language, mannerisms, preferences—indicate that profound social discontinuities still demarcate this transitional zone of Asia. For example, whereas dense population and intensive farming have strictly regimented spatial and temporal practices among the Han Chinese, dispersed settlement and mobile stock-herding have permitted traditional Mongol society to operate and evolve along very different horizons. Such deep-rooted orientations still resist facile alignment with the alien standards now imposed from Beijing, however disguised in the language of economic development.

By appealing to the Great Wall's evident symbolism, I do not mean to portray a simplistic antagonism or rigid historical dichotomy between the settled Chinese cultivators of the plain and the mobile pastoralists of the steppe. A large body of literature makes clear that the two groups interacted with regularity and complexity (see Lattimore 1962; Barfield 1989; Jagchid 1989; Di Cosmo 1994). But although Mongol herders and Han farmers may have lived in perpetual interaction for centuries, the fact remains that they did not live together. Even after military and political subjugation, Mongol grassland communities maintained areas with a distinct space and life-world right into the Communist era. Stereotypes of difference can be overdrawn, but there is no denying that such significant experiential alternatives have remained palpable. Nor is the longevity of difference surprising. As Khazanov (1994: lviii) has noted: "Pastoralism is not only a way of making a living; it is also a way of living, dear to those who practice it."

Despite the eventual penetration of railroads and highways, the heavy stream of in-migrating Han farmers, and increasing contact with the global political economy that new commercial linkages have facilitated, past and present lifestyles continue to clash dramatically without stable resolution. Unrelenting modernization processes exert tremendous pressures upon the social life and identity of contemporary herders who try to cope with the profound changes accelerating around them. Every day, in one situation or another, local residents are challenged to define themselves anew by choosing between social categories that seem to be mutually exclusive—are we Chinese or Mongolian, agricultural or pastoral, sedentary or mobile, privatized or collective, modern or traditional? Such basic questions yield no straightforward response, for familiar meanings have become blurred as global, national, and local influences increasingly mingle in unpredictable ways. Of course, social change and the renewal of meaningful identity have never been unproblematic, but the challenges are now more conspicuous and comprehensive. They intrude upon the daily life-world of every resident.

## Grassland Sketches

Let me quickly reveal a sense of the contemporary life-world of Inner Mongolian herders by introducing just a few of the individuals I got to know over the course of my research. These short biographical accounts accurately portray the kind of routine circumstances and commonplace problems that confronted the people I met and lived among. In some cases I have adopted the use of a pseudonym (here and throughout the book), but all other details are factual.

*Wuliji* is a forty-eight-year-old man who has walked with a cane ever since his legs were maimed in a winter injury in 1993. After drinking among cohorts all day, Wuliji realized he would soon run out of fuel to burn for warmth through the night. Because it was late in winter, he could not expect to scavenge dung patties or willow twigs anywhere nearby, so he sneaked into the locked compound of the local cooperative store to steal some precious coal. He loaded more than he could carry, however, and dropped the heavy bag of coal on his legs. He lay there exposed for some time before a sleeping night guard awoke and discovered him. Embarrassed to be caught asleep on duty, the young guard returned the coal and then tossed the injured intruder outside the compound gate. Wuliji managed to crawl the short distance back to his home on numb and damaged legs before collapsing on top of his smoldering hearth. Minutes later his clothes caught fire and smoke spilled out into the alleys of the village. An alert neighbor rescued him, but not before flames had badly burned his stiff legs. Wuliji never obtained medical attention for his injuries. He had no insurance coverage and was too poor to pay the transportation and medical fees involved in a journey to the regional hospital.

This was not the first time that alcohol intoxication led to disastrous results for Wuliji. Neighbors report that he often lost his temper when drunk and frequently beat his wife at home until she fled the village one day a few years ago, never to return. She left him alone to look after an adult daughter who is mentally retarded and paraplegic. Wuliji now considers whether to send her away to an urban center, where she can beg in the train station to support herself. (Although this plan sounds brutal, it is worth noting that Wuliji's peers evaluate such a course of action against the alternatives of abandonment or outright murder. Indeed, disturbing rumors surfaced during my fieldwork that two brothers had quietly disposed of their own father out in the desert after he became infirm and mentally unstable.)

Wuliji's new physical disabilities severely constrain his own prospects for economic production. In a community still defined by mobile pastoralism, he cannot compete well for scarce resources as an immobile herder. Although he maintains a few cows that wander through the residential center of his village, his small herd dwindles every year. His cows delivered two calves last spring, but he had no land on which to nurture them to maturity. He thus faced the dilemma of watching them starve to death or selling them for a pittance. This situation is not supposed to occur in pastoral communities. By customary law, the village should have a special community-access pasture reserved just for spring calves, but that field was illegally fenced in 1987 by a few elite households with close kinship ties to local government officials. They now control it as private pasture for their own exclusive use. Some residents report that a county-level delegation toured the village in 1989 and ordered those households to dismantle their illegal enclosure, but then left without enforcing their judgment. To this day, the private fences still restrict community access, apparently because the village party secretary is not inclined to direct his authority against the interests of his uncle. Meanwhile, impoverished residents like Wuliji pay enormous social costs for the loss of the common reserve pasture.

*Qiqige* is a forty-one-year-old mother of five children. Local authorities are not happy about the size of her family and have taken a series of punitive measures against her, including recurrent monetary fines, forced sterilization, and threats of imprisonment and confiscation of property—all of which they justify in the name of desert control.

After giving birth to four daughters, Qiqige and her husband still yearned for a son.[1] In 1985, she undertook an expensive pilgrimage to Beijing in order to kneel before the Panchen Lama (a Tibetan spiritual leader) and secretly pray for a male heir. Upon her return, she became pregnant and later delivered a healthy boy. For this breach of family planning, local officials fined her 50 yuan—no small penalty in a village where a person's net income averages only 400 yuan per year—and required her to get her tubes tied. After a botched procedure in 1987, she has become pregnant twice and both times local officials forced her to abort the fetus in a distant hospital at great personal expense.

In 1992, an adjusted family planning penalty structure was implemented, and local officials presented her with a new fine of 3,500 yuan. She was told that a 20 percent discount could be obtained if the money was paid within a week's time, or they would put her husband in jail and confiscate her livestock to satisfy the debt. Qiqige scrambled to borrow the funds from family and friends throughout the region and then raced back home against the deadline. Because of miserable road conditions and unreliable bus service, however, she could only ride to within 25 kilometers of the family planning office, so she covered the remaining distance on foot to make the payment by midnight. She survived the crisis, but she now complains publicly that the latest fines levied against her and others were illegal means to extort money so that local officials could purchase a jeep and motorcycle. She wonders how long it will be until they come around again to impose yet another penalty. She also reports that local officials continue to harass her family through new threats of land seizure.

Since 1991, Qiqige and her husband have worked long and hard to enclose several hectares of moving sand dunes that surround a large pond. They purchased secondhand, inferior fence-wire with cash that represented roughly ten years of family savings. They laid claim to this dismal tract of land because it was fairly close to their mud-brick home and because they had no reason to hope for better. Most all the land near their home was just as bad, and any portion of the range covered by trees or grass had been claimed by more prosperous neighbors long ago.

For years her family poured their energies into nurturing the land to productivity. After leveling a section of dunes by shovel and irrigating them with pond water, they established a nursery of several hundred poplar trees. Just as the area began to show signs of economic potential, the village party secretary came by to announce that the land had been confiscated for government use "in order to combat desertification."

Qiqige's husband resigned himself to the situation and consented without protest. With no political clout in the village, he had no stomach for opposition. For her part, Qiqige was fearful but furious. In her words, "desperation finally pushed me to action." She demanded a hearing with the party secretary and the village chief. Finding a clever rhetorical tactic that was persuasive yet subtly derisive, she presented the officials with an analogy. She argued that her marriage operates on the same basis as the local government. Any household decision by the husband requires sanction by the wife, just as any administrative decision by the party secretary requires sanction by the village chief. By comparing the secretary to an ineffectual and weak husband, she managed both to challenge and insult his unilateral decision to reclaim the land.

She then pointed out that if the government wanted to fight the desert, she was already successfully doing so. They could engage the battle in countless other locations with her blessing. Further, no one had indicated any plans for the land during her years of labor, so it was unjust, she argued, to reclaim it just at the moment of fruition. Finally, she challenged the secretary to confiscate also the land of her more privileged neighbors—people who enjoyed close kinship relations with the secretary. Qiqige asked him, "Why should we alone bear a total loss while others prosper for the same behavior?"

The village officials only consented to give the matter further consideration. In the meantime, Qiqige prepares for the worst. She asked me to take a photograph of her nursery to use as evidence in a future appeal to county-level authorities, should she lose control of the land. When I later presented her with the photo, she accepted it, but pessimistically stated that it would do her no good. She still works the land to preserve her claim on it, but does so "with much less enthusiasm." In fact, she now regrets her strong words to the officials and fears for the safety of her family members—apparently with good reason. Not long after her defiant protest, her younger sister was deliberately hit over the head with a shovel by an irate neighbor furious over the placement of a fence post along a disputed boundary. The sister was hospitalized for three days, and the violent act gained further significance by the fact that the neighbors had once shared close emotional ties and "fictive kin" relations with the victim. Qiqige believes that the sudden attack against her sister was prompted in some way by her own quarrel with village officials.

*Buhe* is a forty-five-year-old male herder with five children and many pressing problems. He has recently been in violent boundary disputes with his older neighbor, Chunbala. Several months ago Chunbala set fence-wire around his homestead but could not afford to complete the job properly. Only the lower line of fence-wire was set, and it was so low that a determined cow could jump across it. This became a problem later in autumn, when fresh forage became more scarce on the range and Buhe's cows spied Chunbala's haystack beyond the fence line. Buhe's children tried to prevent the cows from crossing and immediately retrieved them when they did. But the wandering livestock ate freely during the night. One evening, Chunbala captured a wayward cow and tied it down until Buhe himself came to retrieve it. He then insisted that Buhe pay two carts of hay as compensation. Buhe promised he would deliver the hay as soon as the harvest was complete, but events took an ugly turn the next night when the same cow again ate from the tempting haystack.

Chunbala awoke from his sleep, and in a fury he broke off one of the cow's horns. Buhe had no choice but to slaughter the anguished animal the next day and sell the meat for what he could get at unfavorable market prices. The next night another cow crossed the fence-wire, Chunbala again broke off its horn, and Buhe likewise had to butcher and sell the meat for a substantial loss. These actions set off a bitter feud between the two households that spoiled many years of friendly cooperation. Buhe resolved to enclose his own pasture area and declared his intention to stab to death any living thing that dared wander across from Chunbala's property. Buhe temporarily transferred his livestock to another pasture, and then he established new fence posts for his own line of wire that would cut Chunbala's cornfield in half. This act prompted several more exchanges of hostile threats and vandalism.

After a time, tempers cooled somewhat, and the two households summoned township officials to arbitrate. Five officials arrived and feasted for half a day at Chunbala's home before they measured the distance between both homesteads. They ruled that the fence should run exactly halfway between them, placing the border well outside the cornfield. Both sides agreed to the arbitration and a prominent boundary marker bearing a red flag was erected. Of course, two points define a line, and since only one flag was posted, disputes eventually emerged about the exact contour of the compromise boundary line. Buhe and his wife, Longtang, now complain that Chunbala did not honor the arbitration, beginning his new fence line along the border of his cornfield, but wending increasingly toward their land. They claim that this was his scheme all along, and that he bribed the delegation with a sumptuous meal to forestall any future complaint against his aggressive tactics.

Like Qiqige's family, Buhe and Longtang are financially strapped and vulnerable to administrative disfavor because they also have been fined for exceeding the birth quota. After giving birth to three daughters, they intended to terminate the next pregnancy until a barren sister told Longtang that she wanted to raise the baby as her own. Longtang concealed her pregnancy by hiding in another township, but when she delivered yet another girl, the sister refused to adopt it. At that point, an ethnic Han couple came forward to ask for the child, but Buhe and Longtang would not give it to them. They kept the child for themselves and thus incurred a government fine of 10,000 yuan. Of course they could not pay it, so officials confiscated much of their personal property, and have continued for years to seize the new lambs and calves that are born to their dwindling livestock.

In hopes of providing more revenue for the household, they reluctantly decided to send their teenage daughters out of the township into unfamiliar urban areas to find work in the dangerous brick factories. During my last home visit, Longtang was sick with anxiety for her daughters. She had just received an alarming letter indicating that one had already sustained a hand injury, and that neither had yet received any pay after two months of labor. She wanted to tell them to come home immediately, but she knew of no way to contact them.

These are only three of many disturbing stories I gathered from different villages in a single township. I have many more to tell in subsequent chapters, but these suffice to characterize the rugged life of grassland residents and to illustrate that local land use conflicts are deadly serious, as well as intimately connected to a full range of troubling social issues. These accounts also suggest the prominent role of social power and political fear in shaping economic and ecological outcomes. Wuliji, Qiqige, and Buhe and their families all suffer from various abuses of power played out within their local framework. Subsequent discussion will show how these explicit abuses are compounded by more subtle and nuanced methods of control that also exist within national and international frameworks of action.

## Topomorphic Revolutions and Their Significance

This book is about a pastoral Mongol community caught between a mobile past and a sedentary future, living precariously in the growing insecurities of an unsustainable present. Community anxiety feeds from many sources, but it ultimately derives from two related processes: a declining natural resource base threatened by shifting sand dunes and population pressures, and the disruptive influences of recent government initiatives to "rationalize" animal husbandry production. I will document the behaviors, values, and perceptions of local residents as they respond to one of the most recent and most powerful modernization processes to intrude upon and transform fundamental physical and social realities on Inner Mongolian grasslands. I refer to the momentous impact of the decollectivization and privatization of land use since the early 1980s and, more specifically, to the new production environment that a government-directed enclosure policy has introduced over the last two decades.

Since Deng Xiaoping initiated broad national economic reforms in December 1978, the central government in Beijing has taken action to quicken the replacement of traditional pastoral peoples with commercial livestock producers. Recent policy initiatives attempt to turn an open-range system of grazing into an intensive production regime based on enclosed pastures, irrigated forage production, stall feeding, machinery such as water pumps and tractors, improved breeding, and chemical fertilizer. The first step in the process was to reorganize communal forms of ownership and land tenure. In farming areas, the return to private land use simply reinstated a familiar relationship to property. But in pastoral areas, private land tenure was a radically new and unfamiliar institution for herding households who have traditionally viewed the range as a common property resource.

During the post-Mao reform era, the deceptively simple medium of barbed-wire fencing has launched nothing short of a "topomorphic revolution"[2] in small communities across the northern grasslands. That is to say, a new configuration of public space has set off a chain of transformation with important consequences for both the ecological and the social landscapes of the region. With regard to the physical landscape, private enclosures tend to exacerbate widespread problems of land degradation and ecosystem decline. With regard to the social landscape, enclosures have broadened disparities of economic wealth, leading to new problems of community stratification and fragmentation. Enclosures have also fundamentally altered some of the traditional orientations of space, time, and body for local residents. Such dramatic changes in daily routine and social organization, and the relations between these changes and increasing contact with the global political economy, are the primary subjects of this book.

The proliferation of household enclosures in Inner Mongolia constitutes the final phase of a long historical process in which the central government has tried to sedentarize, and thus more fully control, the mobile herding populations at the nation's periphery. During the Qing Dynasty, China first restricted the movement of Mongol tribes to county level (banner) boundaries. Then, a half-century later, the Maoist era of collectivization forced herding households to root themselves geographically into settled communities. Now the reform era of decollectivization has introduced pastoral Mongols to the full bridle of restricted land use for the first time by fixing each household to a specific plot of land. This book suggests that in terms of spatial discontinuity, the process of decollectivization has been at least as disruptive to local orientations as was collectivization, if not absolutely the greater watershed.

The project of modernization has certainly been a long time coming, both to China and to the Northeast frontier. There have been many phases, with dramatic upturns and downturns along the way. In the early twentieth century, Owen Lattimore detailed the implications of momentous changes in the Northeast that would redefine local patterns of production and marketing, most notably the huge population transfer and the extension of railways. He observed that new rail connections stagnated local exchange for the benefit of urban metropoles, leading to a breakdown of traditional life practices, especially among nomadic peoples (1941: 139). Demographic and economic pressures gradually resulted in a loss of rangeland that forced many herders into farming. Although the desire to settle and harness the pastoral production of mobile herders has occupied the energies of Manchu, Japanese, and Chinese officials over the last century, contemporary grassland enclosures mark a significant escalation, and perhaps a decisive turning point, in the long historical march toward pastoral sedentarization and modernization.

But it is also the aim of this book to address conceptual issues that go far beyond the existential perils and production problems of a single community. The kernel of my argument is that culturally determined and historically informed perceptions of landscape prove to be highly relevant to the way contemporary rangeland policies are both framed in Beijing and contested in the daily practices of local land use. Competing beliefs and values expressed within international, national, and local frames of reference actively shape the process and the interpretations of land transformation in Inner Mongolia. Discourses of development and scientific authority are used to impose and conceal the exercise of state power on marginalized populations, who operate under their own, quite different conceptions of environment, identity, and knowledge. Just as "nationality" or "ethnicity" may be said to exist, a people's sense of place and environmental identity may be said to exist and to influence social action in ways that are worthy of the attention of social science.

Chinese rangeland policy initiatives are informed by a long history of antagonism with the grassland environment and its native inhabitants. For centuries, Chinese literati viewed and described neighboring mobile populations and their homelands in the most disparaging terms. These derogatory Confucian attitudes were only strengthened by Marxist orthodoxy after 1949. The Marx-Lenin-Mao line of political philosophy viewed nomadic pastoralism as an evolutionary dead-end standing in opposition to national progress, scientific rationalism, and economic development. Mainstream Chinese intellectuals in the reform era still consider the land and people to be "in the way" of modernization—obsolete and disposable in their traditional composition. In this regard, Mongol herders may be considered representative of so many traditional life-worlds that have been targeted for eradication in the modern era.

Environmental problems in general, and land degradation in particular, often provide national governments a useful pretext by which to impose modern technologies of surveillance and control. The policies that currently shape reality on the Chinese grasslands can be usefully compared to the strategy of African governments decades ago, when some regional leaders reacted to prolonged drought in the Sahel with Machiavellian satisfaction: "We have to discipline these people, and to control their grazing and their movements. Their liberty is too expensive for us. Their disaster is our opportunity" (Ebrahim Konate, secretary of the Permanent Inter-State Committee for Drought Control in the Sahel, cited in Marnham 1979: 9; and Khazanov 1994: 1).

The imposition of a new spatial discipline is a common component of modernization efforts in China and around the world. Economic development mechanically occurs by destroying prior landscapes and refashioning them into an order that is more efficient for particular (market-oriented) purposes. Roads are built, trees cleared, wetlands drained, common property parcelized, multiple land uses eliminated, settlement patterns reorganized, and everywhere, new boundaries are erected. In all this transformation, individual and group identities undergo reevaluation. Minimally, previous identities and systems of meaning are explicitly contested, if not redefined altogether. During this critical process of reevaluation, the local landscape achieves a heightened capacity to serve as a physical and symbolic arena in which differences among community groups play themselves out in response to the questions: Who are we, and to what environment do we belong?

In Inner Mongolia, the symbolism of the landscape might be expected to play an important role in social processes for several reasons. First, herders, no less than farmers, have a deep cultural predisposition to scrutinize the topographic features of their environment. This derives in part from the material necessities of survival in a mobile pastoral economy. It also derives from a long ideological tradition of *feng shui* (geomancy) that inspires nature-associated symbolism. Second, accelerating ecological degradation and rapid spatial reconfiguration has only intensified the already strong inclination of residents to pay close attention to their physical surroundings. Third, the community is quickly becoming more class-based. Under state communism, a strict egalitarian philosophy reinforced by successive political campaigns to "struggle" local elites prevented rural producers from attaining meaningful stratification in wealth. But in the post-reform era, environmental indicators of household status (such as number of livestock, size of landholdings, productivity of pastures, and quality of dwelling) have attained greater symbolic significance in the absence of well-defined economic hierarchies. Residents clearly monitor the landscape to think about themselves and their social relations. This book will explore all of these factors in greater detail and in relation to broader concerns of social theory.

An abundance of research across many disciplines since the 1980s has drawn attention to the subjective dimensions of land use and resource management. One of the most important conceptual tools in that literature has been the notion of "landscape." Landscape generally refers to the subjective life-space that encompasses both material and symbolic interactions between humans and their dwelling places. Numerous studies show that significant differences do sometimes exist between insider and outsider perceptions of the same physical environment (for example, see Cronon 1983; Dove 1986; McGovern et al. 1988; and Zimmerer 1993). In theory, subjective landscapes extend beyond individuals to create larger nested domains of agreement, such as "vernacular regions" in which particular social identities correspond to localized environmental meanings. These meanings may differ sharply at times, either within or between competing social groupings, and between entire political economies. Environmental preferences and contested interpretations of landscape thus constitute potential boundaries of great political and cultural significance. Landscape preference and sense of place never stand alone, however, but exist in constant juxtaposition to other places and other scales, especially the national and the international scales. For modernizing societies around the world, recent penetrations of a globally structured political economy have brought an unprecedented challenge to local identities and the symbolic cohesion of traditional communities.

As an analytic construct, landscape studies render false the separation, on the one hand, between nature and culture, and on the other, between the material and symbolic domains of human activity so routinely assumed within the dominant political-economy paradigm of most resource management studies. The implication follows that scholarly analysis of environmental change may still have much room for growth. Although political economy has now become a well-established "social factor of degradation" that merits consideration in nearly all contemporary land use studies, cultural considerations do not generally enjoy the same prestige. This ethnography attempts to contribute to a growing literature that demonstrates not only the relevance of culture in resource management, but also its intrinsic inseparability from "the political" and "the economic" in the study of environmental transformation.

Landscape studies also invite new ways of thinking about a wide range of governance and social justice issues, such as how public goods (like natural resources) come to be publicly defined, negotiated, and used across a diverse body politic. One specific issue is the popular assumption that common property resources, by definition, cannot be well managed. Since the late 1960s, international policy circles have promoted privatization and/or state regulation to prevent the perceived inevitability of the so-called tragedy of the commons. But my fieldwork and data indicate that the privatization of land in modernizing pastoral societies can be less meaningful for good resource management than other factors such as secure tenure, equitable access to community resources, and meaningful institutional supports in the form of credit, production services, and legal protection. Another governance issue raised by landscape studies, one even more sweeping in implication, is the prospect that ideologically imposed perceptions of the environment—just like ideologically imposed perceptions of race, class, or gender—constitute a significant dimension of social manipulation and oppression, even though we are not yet used to thinking about it in such serious terms (see Dove 1998). Resource management thus constitutes a critical arena in which conflicting worldviews routinely struggle for expression and dominance.

Within this intellectual context, I believe the topomorphic revolution under way in Inner Mongolia provides an excellent opportunity to investigate some of the dynamics of subjective landscapes—to examine whether differential perceptions exist, whether they have any significant implications for resource management at the household level, and whether they demarcate any symbolic communication between social groups with wider social implications. As Smil (1987: 222) once wrote, "If one seeks confirmation for a thesis about social rather than natural causes of land degradation and about the importance of social as well as natural sciences in the understanding of this process, then China is the classic example."

Furthermore, the situation in Inner Mongolia offers an opportunity to investigate some of the more serious unintended consequences of decollectivization that often get swept under the rug of nationalist rhetoric. China's government in particular has been averse to acknowledge and confront the many unpleasant by-products of rapid economic growth for entire segments of the population. A village-level ethnography can help reveal some of the more subtle trade-offs that occur in the development process. More broadly, this research setting offers an important opportunity to increase our understanding of some of the troubling issues that now challenge humanity everywhere, including acute environmental reconstruction, rapid technological and social change, and debilitating crises of identity.

Thus, as a window onto pastoral China, this book reveals some interesting and previously unreported facts about contemporary resource management and living conditions in the northern grasslands. As a window onto human-environment relations, it documents some of the hidden complexities of land use decisions and demonstrates the creativity with which people use ecological crisis to service their own social identity and cultural renewal.

## Why Study Pastoral China?

A village-level case study of land use from pastoral China is long overdue for several important reasons. First, China has a huge pastoral industry of geopolitical significance, though scholars do not often realize it. Dwarfed by the numbers and political centrality of ethnic Han cultivators, minority pastoralists in China remain obscure, on the fringes of Chinese geography, scholarship, and national economic priorities. Yet China has 260 pastoral counties accounting for 39 million people. It has some 400 million hectares of grassland, constituting 42 percent of its land mass. It is the world's third largest grassland, but it supports the world's largest combined population of sheep and goats, and the fourth largest concentration of cattle (Zhang Xinshi 1992: 47). Inner Mongolia itself constitutes fully one-fourth of China's total rangeland area, and in recent years it has been the nation's leading producer of wool, cashmere, and camel hair.

Second, there is little published ethnographic data about contemporary life and land use in the grasslands. The lack of information available in English can be explained by problems of long-term access to sensitive border regions that have only relaxed with the recent collapse of the Soviet Union. The need for information from this area only grows stronger with China's own reassessment of the importance of the region for continued national development. Within a domestic agenda, Chinese leaders hope to increase the dietary intake of protein among its citizens by eating more red meat. They also hope to secure more fully the strategic border regions populated by impoverished minority ethnicities by raising low standards of living. Within an international agenda, Chinese leaders hope to expand exports not only in meat and leather products but also in light industry, which relies heavily on raw materials provided by sheep and goat husbandry.

Third, Chinese scholarship is not likely to provide essential ethnographic data from this region anytime soon. Chinese grassland research and data collection overwhelmingly focus upon biotic interactions among soils, plants, and herbivores, with little attention to the actual behaviors and motives of human grassland inhabitants. This is particularly problematic considering the nearly universal consensus attributing desert expansion in China to human behavior. When Chinese data do focus on grassland residents, the information is typically limited to narrow economic parameters, reporting such figures as animal units, stocking ratios, and production/consumption levels.[3] A prominent social scientist from Beijing University recently conceded that Chinese grassland studies have generally ignored such basic social concerns as household production strategies, risk management, allocation of labor, and the role of competitive markets and prices.[4] Typically, domestic reports about grassland modernization efforts present social changes casually, as if they pose no particular challenge or disturbance to local residents and community power relations. One news report, for example, stated that minority herdsmen recently discarded their nomadic lifestyle and eagerly enclosed pasture land "without a second thought" (Xinhua 1989). My fieldwork experiences convince me that something far more complicated and interesting is happening on the grasslands.

Fourth, an acute environmental problem now prevails on China's northern rangelands that properly deserves the attention of the world. Soil erosion and desert expansion have seriously limited prospects for sustained agro-pastoral production. Official reports from China routinely assert rather alarming figures: fertile grassland is now lost to moving sand at an average rate of 2,460 km² per year (Xu Youfang 1997), while moving dunes purportedly pose an economic "menace" to as much as one-third of the entire national land mass (Xu Youfang 1993). The situation is said to be particularly severe in Inner Mongolia, where some officials report grassland loss at a rate more than double the national average, or about 6,000 km² per year.

The story has apparently become more newsworthy in recent years, as the scale of disaster expands to affect people at China's political center. The national media often portray Inner Mongolia as "Beijing's backyard" and attribute the appearance of frequent "mud rain" and dust storms in the capital city to the improper use of pasture land in nearby grasslands (Xinhua 2000a). For example, in the spring of 2000, Beijing was hit by a series of twelve dust storms that delayed flights at the inter-

national airport and imposed unexpected economic and social losses on
the residents of the metropolitan area. In response, government officials
have emphasized that it is "high time to deal with man-made deserts,"
and more than $2.4 billion will be allocated over the next decade in an
attempt to insulate Beijing from the ecological hazards widely associated
with Inner Mongolian herders (Xinhua 2000a, 2000b).

## Research Framework and Host Institution

Environmental concern has provided the context in which a variety of
multinational and multidisciplinary research efforts have been organized
on the Mongolian steppe—involving field studies in Mongolia, Russia,
and China—since the 1990s. Asian scientists and officials, anxious to
raise agricultural productivity in the grasslands, have opened up to Western natural and social scientists in a number of large and small collaborative projects. The most ambitious has been the University of Cambridge
MacArthur ECCIA (Environmental and Cultural Conservation in Inner
Asia) project, which involves both social and environmental research in
the comparative study of the Inner Asian steppe and its peoples across
Mongolia, Inner Mongolia, Buryatia, Tuva, and Xinjiang. Much of this
research has been published under the names of Caroline Humphrey and
David Sneath (Humphrey and Sneath 1996a; 1996b; 1999).

Another initiative in Mongolia is the Policy Alternatives for Livestock
Development Project (PALD), based at the Institute of Development
Studies in Sussex. Much of this research has been published under the
names of Jeremy Swift (1993) and Robin Mearns (1993a; 1993b). The
Nordic Institute of Asian Studies in Copenhagen organized a Central
Asia workshop in 1993 that examined continuity and change in Mongo-
lia and produced a collection of research papers edited by Ole Bruun and
Ole Odgaard (1996). Another project, focusing upon Inner Mongolia,
Xinjiang, and Gansu province in China, brought together Australian and
Chinese agricultural economists to study pastoral wool production and
marketing. This work was funded by the Australian Center for Interna-
tional Agricultural Research (ACIAR), and has produced publications
under the names of John Longworth and Greg Williamson (1990; 1993).

In the United States, a loose organization of scholars identified by the
acronym GEMS (Grassland Ecosystem of the Mongolian Steppe) was
nurtured by the Committee on Scholarly Communication with China under the leadership of James Reardon-Anderson. Following upon the publication of a volume that surveyed the general state of grasslands and

grassland sciences in China (NRC 1992), the GEMS project organized two working conferences and attempted to facilitate inter-disciplinary research networks among Mongolian, Chinese, European, and American scholars. The forum was inclusive, so that scholars involved with both ECCIA and PALD also participated in GEMS.

The GEMS project was instrumental in placing me in a minority pastoral village where I could conduct unchaperoned social research under the protective sponsorship of the Shenyang Institute of Applied Ecology within the Chinese Academy of Sciences, which operates a small weather station and grassland research outpost in a township that lies within China's easternmost desert area.[5] The scientists from Shenyang have controlled a sizable portion of enclosed land (about 3,000 hectares) throughout the township since 1970 in order to conduct experiments in dune fixation and afforestation. Over the last twenty-five years, a dozen or so Han scientists[6] have rotated in and out of the area, usually in small groups, for a month at a time from March through October. Their self-described mission is to study and disseminate information about comprehensive grassland management strategies, to improve and intensify the rational use of local resources, and to raise animal husbandry production capacity. A billboard in the research station conference room defines the general scientific mission in rather blunt terms: "to change the natural appearance of the area."

Scientists at the Shenyang Institute invited me (as a participant in GEMS) to live for one year at the research station in order to investigate the human and historical dimensions of grassland degradation in the area. The context of resource management thus allowed me to gain long-term access to an otherwise restricted and relatively uninvestigated area of China. The unusual juxtaposition between Han scientist and Mongol herder also afforded me a convenient opportunity to explore differential group perceptions and attitudes toward local landscape ecology, as well as toward economic development interventions and the ongoing work of the scientists.[7]

The research station originally acquired its land base through negotiations with local commune and brigade leaders. It was reported to me that brigade work units assisted the scientists in bulldozing land and constructing the various living and research facilities, including dormitories, a kitchen, a garage, offices, a conference room, a laboratory for soil-plant analysis, a weather station to record atmospheric data, a shelter-belt and nursery to surround the compound, outlying artificial meadow and for-

est areas for controlled experimentation, and additional scattered plots
for the study of dune fixation, alkalization, irrigation, vineyard and or-
chard production, and even wet-rice cultivation. Throughout the region,
the scientists have enclosed (but do not monitor on a daily basis) an estimated 40,000 hectares of experimental plots (Zhao Shidong 1992: 2).
Supported by a series of five-year research grants, the scientists understand their first priority to be data collection and experimentation, and
their second priority to be the demonstration of scientific techniques to
local farmers and herders.[8]

Over the years, the scientists have recruited a small network of
"model" households who are willing to receive instruction and assistance
in planting trees, treating soil, cultivating fodder crops, or developing orchards. Yet the relationship between the research staff and the community has been problematic, vacillating between the norm of passive acceptance and the extreme of open hostility. Located in the heart of the
residential center of Wulanaodu, the station has isolated itself spatially
from the community by constructing high, thick walls all around the
compound. In my year of residence, the staff maintained a working relationship with just a few key residents, largely going about their business
without actively engaging the native population. Some residents take advantage of opportunities for temporary employment provided by the station, others appreciate random favors such as free transportation between villages, and some will even assert that the township has derived
important benefits from the scientists, especially from earlier forestry
projects. Many more residents, however, openly grumble about the confiscation of so much land and the strict prohibition against both grazing
and hay production in these areas. For their part, the scientists claim that
they only enclosed the most desolate tracts of land for observation and
experimentation, which is essential to plan the restoration of the entire
ecosystem. They insist that most of the land they enclose must lie absolutely fallow to serve as a valid control.

## Doing Fieldwork in Nasihan

I conducted anthropological fieldwork over a period of twelve months
during 1993–94 in Nasihan township (*sumu*) of Wengniute banner (*qi*),
Chifeng City prefecture (*shi*) (see Maps 1.1 and 1.2).[9] I engaged in par-
ticipant-observation and collected data throughout the entire township,
but the bulk of my contact occurred with residents in my host village of
Wulanaodu, where I enjoyed relatively unimpeded access to government
documents, household registries, and personal interviews. In 1993, the

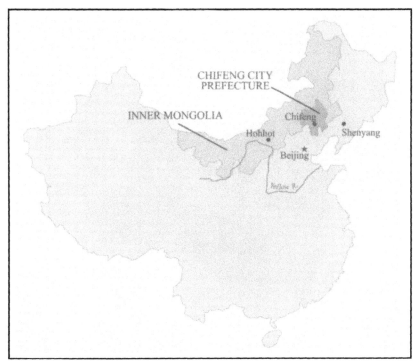

MAP I.I. People's Republic of China, with IMAR and Chifeng City Prefecture

herding township of Nasihan had a total population of nearly 4,000, with 95 percent comprising ethnic Mongols. Wulanaodu, the largest of ten township villages, had a population of 740 people divided among 174 families, and a population fully 98 percent Mongol.

The population speaks both Mongolian and Mandarin Chinese, but I communicated with residents in Mandarin only. For those few who could not converse in Mandarin, I relied upon the translations provided by household members or helpful neighbors. Stock-herding of cows, sheep, goats, horses and camels still accounts for 87 percent of village income. According to local government statistics, the total village herd in the summer of 1993 consisted of 1,829 cattle, 1,270 sheep, 4,012 goats, 248 horses, 81 donkeys, and 64 camels. The average number of each animal species per active herding household is roughly 11 cows, 8 sheep, 25 goats, 1.5 horses, 0.5 donkeys, and 0.4 camels, although not every household always keeps every species. The cash economy is small, with a per capita net average income of only 400 yuan per year (about U.S.$50). The area thus ranks among the poorest townships in all of China.

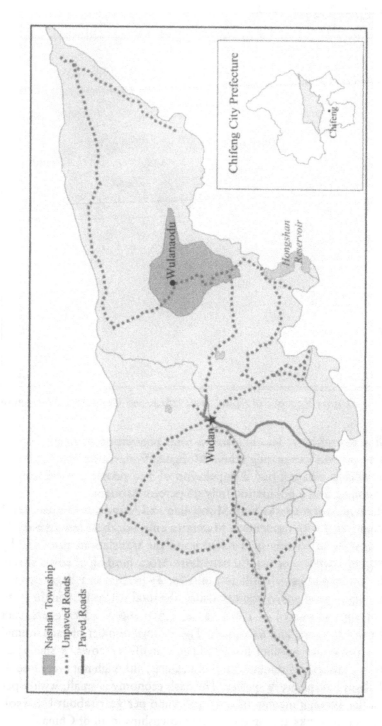

MAP 1.2. Wengniute banner, with Nasihan township and Wulanaodu village

Chifeng City Prefecture

Chifeng

Wulanaodu

Hongshan Reservoir

Wudan

Nasihan Township
Unpaved Roads
Paved Roads

Nasihan township is situated in a desert-steppe environment in the western portion of the Keerqin Sandy Lands. The total rangeland area is roughly 613 km², but officials co sider 97 percent of it to be "deteriorated" (Nasihan 1991: 1). The clim te is semi-arid, with mean annual precipitation usually ranging between 300 and 500 mm. The yearly mean rainfall from 1957 to 1990, as recorded in the banner seat of Wudan, amounted to only 368.8 mm. There is a strong seasonal pattern to local rainfall, with 71 percent of mean annual precipitation occurring during the summer months of June, July, and August.

Although my living conditions were frequently spartan (no running water, no electricity, no nutritional variety, no health facilities), I was joined in the field by my wife and one-year-old daughter for six of the twelve months of research. They arrived later than I and retreated to a safe environment from the end of October until the first of April. I also left the village during the most brutal weather, from December through February. Indeed, I could not have survived the winter on location because I was not equipped with the minimum necessities that every native household uses to survive: a stove and *kang* for heat,[10] abundant quantities of stockpiled food for sustenance, and a large network of friends and family to cope with the loneliness. The research station itself does not operate during winter months.

I should point out briefly that my wife and child had an important effect upon my research. In addition to the benefit of their company, they helped me gain and maintain friendly ingress into the community. After they arrived, village residents suddenly perceived me as much more interesting and likable. I also suddenly found it easier to discuss family and domestic issues. Furthermore, female perspectives began to surface in my research, not only because I could talk to women more easily, but also because my wife (who also speaks Mandarin) eventually developed her own network of female friends. As part of an American family, I was warmly received; as a foreign anthropologist, I was often actively avoided.

I would characterize my relationship with the scientists and staff at the research station as cooperative and friendly (even jovial most of the time). My daily interactions with them blurred the lines of work/play and visitor/host, so that I quickly graduated from the status of "guest" to that of "colleague" and even "confidant" at times. Our living circumstances were simply too intimate and difficult, and my duration of stay just too long, to maintain the conventions of polite separation that normally characterize Chinese relations with foreign delegations. (Indeed, since the scientists rotated in and out of the station while I remained behind to send them away and welcome them back, we frequently joked that I had become the true "host" of the compound.) Of course, we also shared many significant interests and experiences that bound us together both professionally and emotionally.[11]

Over time I also developed a positive working relationship with the local residents, but it was a long and difficult transition. The scientists never asked the residents for permission to invite me (nor anyone else) into their community. I was introduced to local administrators long after the fact of my arrival, and I only gradually found ingress into village life through my own persistent efforts. At first, residents were understandably quite hostile to my presence. Many of them assumed that I was a spy for the American government, although nobody could imagine what I might have come to investigate. Thus, the relevance of social power to my very presence in the village was explicitly recognized by all. Given this circumstance, I felt great achievement in the eventual success of gaining broad acceptance and even emotional attachment among so many households by the time of my departure.

The remote and impoverished setting made for extremely arduous fieldwork. To complete my household interviews, I had to travel across mobile dunes into far distant pastures in search of isolated homesteads that are almost always guarded by ferocious dogs. I owned a motorcycle, but many home visits could only be accessed by mounting on horseback, riding in a mule litter, or marching cross-country. Because the journey itself would consume so much daylight and bodily energy, I often relied upon the generous hospitality of the herders to eat, drink, sleep, and find my way from place to place. A night stay almost always involved obligatory consumption of alcohol and sleeping together with the entire family on a large *kang*. This circumstance put me in a good position to observe in a fairly intimate way the various living conditions and routine behaviors of the people. Of course, it also set me up for a number of amusing, distressing, and sometimes even dangerous experiences,[12] all of which served to help me gain further insight into the community.

Finally, it is important to note that Nasihan township, and Wulanaodu village in particular, are places where the proliferation of private household enclosures is relatively advanced within China. This fact follows from the advanced state of land degradation, as well as from other unique historical circumstances. In 1958, for example, the village received national recognition as a model pastoral commune. Residents report that

Mao himself praised the village in a speech from Beijing, and he ceremoniously awarded it a symbolic red star of merit. The fame later served to gain the collective certain advantages, such as subsidized fence-wire for the purpose of enclosing reserve meadows for winter hay production. It was in large part a result of this subsidized wire that elite households had the early opportunity upon decollectivization to confiscate public assets and assert their own private pastures. The wider region of Wengniute was also targeted through the early 1980s for special production assistance by international development institutions. For these reasons, Nasihan experiences with the enclosure movement anticipate somewhat the transformations that are just beginning, or yet to occur, in other pastoral areas of China.

# TWO

# Land Degradation and the Chinese Discourse

China has eight distinct gravel desert zones to which the Mongol term *gobi* is applied, and four sandy desert zones to which the Chinese terms *shadi* or *shamo* are applied. Gobi differ from shadi in several respects: they consist primarily of stony or gravel deposits, they lie to the west (windward side) of the steppe zone, and their dunes are more mobile than the semifixed or fixed dunes characteristic of the east (Zhao and Xing 1984: 230). Primarily as a result of strong wind transport, the soils of arid northern China—moving across the grassland from northwest to southeast—generally follow a progressive transition from gravel to sand to loess (Fullen and Mitchell 1991: 26).

The eight gobi regions, accounting for roughly 42 percent of China's total desert area, include Taklimakan, Gurban Tunggut, Kumtag, Qaidam Basin, Badain Jaran, Tengger, Ulan Buh, and Qubqi (Hobq). The sandy lands, accounting for roughly 58 percent of China's deserts, include Mu Us, Hunshandak (Otindag), Keerqin (Horqin), and Hulun Buir. Together, these areas link into a sand belt that stretches some 5,000 km from west to east across the northern provinces. The belt extends over the autonomous regions of Xinjiang, Ningxia, and Inner Mongolia, and over the provinces of Qinghai, Shaanxi, Liaoning, Jilin, and Heilongjiang. (Map 2.1 presents a schematic view of China's northern deserts.) The sand belt, however, does not lie inert. Its boundaries change over time, sometimes quite dramatically. Of course, the once popular and foreboding notion of "desert creep" has been replaced in recent scientific literature by a more nuanced and complex imagery depicting pockets of deterioration that eventually enlarge and merge (see Heathcote 1983; Nelson 1990). Nonetheless, desert areas are known to be dynamic and may expand (or contract) over time.

At a national level, desert expansion has generated a great deal of government concern and public anxiety. Prominent officials now estimate that grassland is lost to moving sand at a rate of 2,460 km² per year (Xu Youfang 1997; Xinhua 2000c), compared with a rate of 2,100 km² per year throughout the 1980s, and a rate of 1,560 km² per year throughout the 1970s (China State Council 1994: 181). They now classify 27.3 percent of the national land mass as desert area (Xinhua 2000c), compared with 15.9 percent reported in 1993 (Xu Youfang 1993). Desert expansion purportedly affects the livelihood of nearly 400 million people and causes direct economic losses estimated at more than U.S.$3 billion annually (Wang, Wang, and Zhang 1993: 1; China State Council 1994: 180). Although official figures tend to vary inexplicably from one source to the next, all domestic reports do seem to agree on the fundamental premise of an accelerating ecological crisis.

In Inner Mongolia, a high government official reported in 1993 that regional deserts were expanding at a rate of 3,400 km² per year (Zhou Weidi 1993). Of an estimated 86.7 million hectares of grassland (nearly 70 percent of the total land area), officials consider 34.5 percent to be deteriorated, and 21.6 percent to be seriously deteriorated or unusable. That leaves, at most, only 43.9 percent in decent usable condition (NRC 1992: 18), while some estimates put the figure as low as 32 percent (see Zhou Weidi 1993). Furthermore, officials estimate that since 1965, total grass production has declined by 30 percent (NRC 1992: 18; *Neimenggu ribao* 1990). On average, each hectare of land produces only about 750 kilograms of haystraw per year, though the range of edible offtake fluctuates tremendously from pasture to pasture (Longworth and Williamson 1993: 81).[1]

Within Wengniute banner of Chifeng City prefecture, animal husbandry officials report that roughly 603,000 hectares of grassland, or 87 percent of the total rangeland, is now deteriorated. That figure is up from 413,000 hectares in 1965, and 493,000 hectares in 1976 (Longworth and Williamson 1993: 188). Within Nasihan township of Wengniute banner, sand or moving dunes occupy 90 percent of the land, and officials estimate that only 51 percent remains at least marginally useful for livestock production. The most fertile pastures were enclosed in the 1960s under collective authority; this was done to reserve land for autumn hay production in order to tide household animals through the long winter months. This vital area, however, occupies only 3 percent of the total rangeland, and yields from even this limited area are said to be declining yearly.

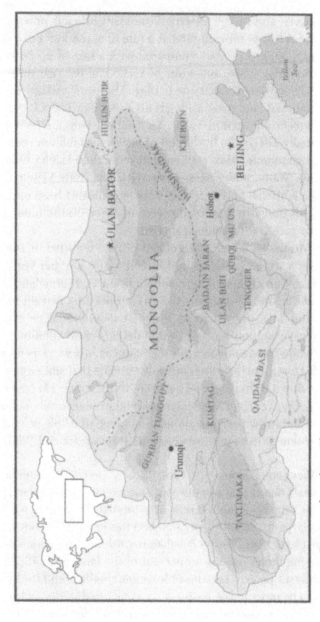

MAP 2.1. Deserts of northern China

TABLE 2.1

*IMAR Land Quality Estimates at Provincial, Banner, and Township Levels*

(% of total land area)

| | IMAR | Wengniute banner | Nasihan township |
|---|---|---|---|
| "Unusable" (a) | 22 | 9 | 46 |
| "Deteriorated" (b) | 34 | 78 | 51 |
| SUBTOTAL (a + b) | 56 | 87 | 97 |
| Remaining "Good Pasture" | 44 | 13 | 3 |

SOURCES: NRC 1992: 18; Longworth and Williamson 1993: 188; Nasihan sumu official document 1991.

Table 2.1 summarizes the reported figures of land quality at the township, banner, and provincial levels. These numbers indicate that the problem of desert control is especially acute in the field site of Nasihan.

## Conventional Explanations of Desert Expansion

Chinese government officials, scientists, and scholars widely attribute the cause of land degradation and desert expansion to past and present anthropogenic forces. Though climatic and physical processes first formed the deserts of China, humans have contributed to their enlargement. Officials within the Ministry of Forestry have estimated that only 500,000 km² (or one-third) of the current total desert area was formed by nature—"the rest has been the making of human activities" (Bureau of the Ministry of Forestry 1990: 22). One Chinese scholar contends that in the Ordos region, the rate of desert expansion owing to human factors since the 1960s exceeds the natural rate of expansion over the previous 2,000 years (He 1991: 24). According to Zhu Zhenda, one of the foremost authorities on the subject, sand dune encroachment by natural causes accounts for only a tiny percentage of the current ecological problem. He asserts that "only 5.5 percent of lands of desertification results from invading sand dunes; the great majority, 94.5 percent, may be described as having undergone desertification in situ initiated by human activities" (1990: 70).

The causes for land degradation in Inner Mongolia are especially attributed to anthropogenic pressures. Explanations of human impact usually begin with an account of exponential population growth: "too many people and too many animals are pressing too hard on a fragile ecosystem" (NRC 1992: 33). In Wengniute banner, officials begin any discussion of rangeland management by pointing to "excessive" numbers of people and livestock. Without a doubt, there have been significant demographic shifts over the twentieth century that deserve a brief summary.

## POPULATION GROWTH

The integration of Inner Mongolia into a single political entity of China proper occurred during the Qing Dynasty (1644–1911), when both territories were conquered by neighboring Manchus. In the closing decades of Qing rule, government officials permitted the settlement of Han farmers into the grasslands, and finally even encouraged it. They were eager to alleviate mounting political instabilities that resulted in part from widespread famine and desire for land. Incrementally, Han colonization expanded across traditional Mongol rangeland. The influx intensified after 1911, when the new Chinese Republic declared that all Mongol lands belonged to China and that land titles were henceforth invalid unless ratified by local Chinese authorities (Lattimore 1934: 105; Jones 1949: 61). By 1924, when the railway line was extended from modern-day Zhangjiakou to Hohhot and Baotou, Han settlers immigrated by the millions, scattering Mongols from their most fertile grazing pasture. The population of Inner Mongolia in 1912 was roughly 2.04 million, with a ratio of 1.3 Han to every Mongol (Ma 1984: 111). By 1990, the total population rose to 21 million, with a ratio of 6 Han to every Mongol.[2] In Chifeng City prefecture, the numbers tell a similar story. The population of 1912 totaled about 700,000, with 1.37 Han to every Mongol resident. By the beginning of the reform era in 1979, the population reached 3.51 million, with nearly 11 Han to every Mongol (Ma 1984: 111).

According to nearly all historical accounts, the large-scale changes in land use and the increases in demographic pressures associated with Han colonization did escalate ecological changes within the steppe zone. During the 1930s, for example, Lattimore was attentive to the problems caused by migrating settlers who had no experience handling livestock and used the land in the only way they knew how—cultivating it, despite inadequate rainfall and unrelenting wind force. He wrote:

The type of colonization created by the rapid building of railways demanded quantity rather than quality . . . [because] no supply of colonists with capital of their own was available. Consequently the land came under the control of capitalists who could afford to take over large holdings and place tenants on them. The colonists had no experience in handling livestock. . . . In order to produce financial results, land had to be farmed even if it was naturally more suitable for grazing than for ploughing. The good soil is then blown away, and sand begins to work up from below. . . . Such districts become totally unproductive, for even if they are abandoned, the old growth of grass will not come back; at least not for many, many years. Human action is rapidly extending the desert areas in Mongolia. (1962: 421–422)[3]

He also described the implications of agricultural extension for areas that remained pastoral:

Pastures have become overcrowded, and the decrease in real nomadism means that herds are kept too long on the same pastures, with the result that the pastures become "stale" and the herds less fertile and more subject to cattle plagues; while the overcrowding of sheep and goats, whose sharp hoofs cut the turf, has a ruinous effect in destroying the topsoil and creating first erosion and then sand dunes that is little less wasteful than the agriculture of Chinese colonists. (ibid.: 446)

Mounting demographic pressures have not abated since the founding of the People's Republic. Since 1950, the vegetative yield of China's grasslands has shrunk by half while the number of livestock has quadrupled (Hinton 1990: 84). Throughout Wengniute banner, symptoms of overgrazing have appeared since the 1960s, with declining pasture yields increasingly manifested in a "marked decline in animal yields and an increase in mortalities despite the adoption of improved livestock breeds" (Brown and Longworth 1992: 1666; Longworth and Williamson 1993: 188). In Nasihan itself, the human population has more than doubled, from 1,728 in 1958 to 3,957 in early 1993, while the animal population (in sheep equivalent units) has fluctuated, rising from 64,432 units in 1959 to a high of 95,358 in 1965, then shifting downward again to 65,467 by 1992 (Nasihan sumu official document 1993).

SHRINKING LAND BASE

While human population pressures have consistently increased, the land base available for extensive livestock herding has shrunk significantly. First, the historic decision to expand cultivation in pastoral areas during the national campaigns of the Great Leap Forward and the Cultural Revolution greatly reduced the productive rangeland available to minority pastoralists. Calling for maximum local self-sufficiency in cereal production in the wake of the Great Leap Famine, Mao encouraged farmers to plow up pasture land that was unsuitable to dryland agriculture. High-ranking Mongol leaders in IMAR who denounced this campaign in favor of livestock production were arrested or demoted

(Jankowiak 1988: 272; Sneath 1994: 419). Foreign environmental analysts now believe this disastrous campaign sharply accelerated the degradation of China's farm soils, grasslands, forests, and wetlands (Smil
1987: 216). The scars of failed agriculture and loss of tree cover left the
earth susceptible to strong winds that both remove organic matter and
transport sand. Regional statistics indicate that 21 percent of the total
rangeland was lost to agricultural production between 1953 and 1979
(Longworth and Williamson 1993: 305). This figure reflects the pace of
change that occurred at a national scale during the Mao era, when an estimated sixty-seven million hectares of high-quality rangeland were converted to grain cultivation, while only eight million hectares of grassland
were reclaimed (NRC 1992: 48).[4]

Second, processes of urbanization and the expansion of nonagricul-
tural rural activities in the reform era have contributed to what is now regarded as a serious decline in fertile soil all over China (Orleans 1991;
Howard 1988: 57; Hinton 1990: 74). Rural arable land is increasingly
lost to housing, roads, factories, and grave sites. The declining land base
not only intensifies production pressures on plots that remain under cultivation but also transfers those mounting pressures onto lands of more
marginal quality, generally toward the periphery of agrarian areas where
minority populations reside.

Third, various processes of land degradation have obviously eliminated large tracts of usable pasture. Soil erosion sets up a positive feedback loop whereby the continuous loss of good soil only intensifies production pressures on the remaining areas, so that they become degraded
as well. The end result is that the numbers of grazing livestock exceed the
sustainable stocking rate almost everywhere (Longworth and Williamson
1993: 333).

## The Chinese Official Discourse

Chinese officials try to deflect responsibility for environmental disaster away from anyone associated with the current regime of reformers.
This is accomplished by diverting blame either in space or in time. The
space-oriented strategy places blame on local land users far from Beijing,
who are routinely portrayed as ignorant, irrational, backward, and un-
cooperative. The temporal strategy lays responsibility at the feet of previous governmental regimes, especially the Qing, the Nationalists, and
Maoist zealots.

## BLAME THE LOCALS

Chinese officials and scholars primarily blame local residents for problems of land degradation. They often repeat a standard assertion: "Over-cultivation and a surplus of stock in the region are the main causes for the rapid desert expansion" (Xinhua 2000d). A renowned scholar at the Department of Desert Research (DDR) in Lanzhou stated that urban industry and the state are responsible for only 9 percent of desert expansion, while rural peasants are accountable for as much as 85.5 percent of the national problem (Zhu Zhenda 1990: 70). Likewise, in the high-profile document known as Agenda 21, the China State Council (1994: 181) asserted that "the formation of desertification in China is the results (sic) of over-cultivation, overgrazing, and destruction of vegetation." This statement, designed for international consumption, merely reiterated a conventional formula that pervades Chinese scholarship (e.g., Fei 1984; Zhu and Wang 1990; Hu 1990; Kou and Xue 1990).

Chinese officials and scholars often point to the "ignorance" and "backwardness" of minority peoples. In particular, Mongol herders are widely criticized for holding to traditional, "rely on heaven" (*kaotian fangmu*) methods of production. Environmental restoration, it is believed, can begin only once traditional practices have been abolished: "The traditional pasture system that relied entirely on 'Heaven' should be abandoned. Sophisticated farming techniques should be employed to improve pastureland and to cultivate supplementary feedstuffs. . . . In short, economic development and environmental quality will change to a higher and higher standard" (Zhao Li 1990: 270).

Influential figures in China such as Li Yutang, Guo Yang, Xu and Qiu, and Zhao Zhidong, to name a few, argue that traditional Mongol herders have never concerned themselves with grassland preservation under the mobile conditions of their past. They have never learned to look beyond their sheep to the soil, the theory goes, so today they have no regard for the land that farmers have long cherished (Guo 1993; Li Yutang 1992). The following printed statement is representative: "The core of reform in the grasslands must be to introduce a kind of contract responsibility system which would increase the worth of the land in the eyes of those who live on it, and persuade them to protect the grasslands by convincing them that the grass is their living, as well as their fodder" (Xu and Qiu 1995). Han scientists working in pastoral areas sometimes endorse this crude argument directly: "Lack of development in the area is due to deterioration of the ecological environment, a lower level of culture, technique and productivity" (Zhao Shidong 1992: 2). They may also endorse it indirectly by appealing to common knowledge: "Pastoralists are often said to have little understanding of the delicate ecological balance of the pasture land. As a result of this ignorance they allow their pastures to become overgrazed" (Lin 1990: 88). Sometimes they endorse it even while evoking empathy: "People in these areas don't have a strong awareness of environmental protection and they are also economically underdeveloped. Some don't even have coal to make fires and they just cut the grass to make do instead, thus turning grassland into desert" (*South China Morning Post*, 1994).

Even grassland regulations and policy statements employ language in a way that subtly perpetuates a condescending perspective. For example, they explicitly call upon household contractors to pursue principles of scientific planning (*kexue huafen*), energetic construction (*dali jianshe*), vigorous protection (*jiji baohu*), and rational utilization (*heli shiyong*) (see Chifengshi caoyuan jianlisuo 1990: 7–8; Wengniuteqi renmin zhengfu 1988: 1). Such exhortations are based upon the premise that principles of conservation and initiative are basically absent among minority groups.

A second common criticism is that minority herders are lazy. This aspect of public discourse has been captured and essentialized by the phrase *jin mao dong* (to eat the winter), and it especially raises the hackles of residents in Inner Mongolia. The term is frequently used among friends to refer to the production slack time during winter months. When neighbors greet and inquire after one another, for example, a response of "jin mao dong" indicates no special news. Sensitivity to the phrase apparently arose in the early 1970s, when an assistant to Zhou Enlai delivered a speech in North China in which he referred to it disparagingly, indicating that it condoned slothful inaction. He suggested that the region's major production problem was the laziness of the local inhabitants, who would rather lie inactive during the winter than explore ways to boost productivity. Of course, hard work does continue throughout the winter months, but it mostly involves the routine chores of survival: cutting wood, collecting dung, drawing water, and sheltering animals. Residents do not think it reasonable to expect more than this in such a hostile environment, given current levels of technology and economic opportunity.

Suggestions of laziness also appear in official discourse through other, less provocative phrases. For example, the 1993 annual report of the Wengniute banner government cited both "ideological conceptions that still have not adapted to the requirements of new styles of development" (*sixiang guannian hai bu shiying xin xingshi fazhan de yaoqiu*) and "a passive attitude lacking initiative in thought and action" (*quefa ganxiang, gangan de shouchuang jingshen*) as two of the greatest problems facing regional development (Wengniuteqi renmin zhengfu 1993: 7).

The national media reinforce this discourse by routinely accusing local land users of environmental mismanagement. The following public pronouncement on land degradation by a senior official in the Ministry of Forestry is typical: "The cause is mainly human sabotage. Excessive grazing, rampant cultivation, unchecked digging up of herbs, and misuse of water and land resources have been major factors leading to desertification" (Xu Youfang 1993). The notion of sabotage (*pohuai*)[5] resonates with other formulaic explanations of land degradation that emphasize a mean-spirited and wanton assault on national assets by peasants.[6]

BLAME PREVIOUS REGIMES

Official discourse also sometimes deflects criticism for ecological decline by removing the problem to an earlier time. Authorities point to the long history of resource abuse and neglect along the national frontier and blame previous regimes for aggravating or ignoring the situation.

*Absolve the Maoist era.* Considering the history of colonization in IMAR, many officials and scholars since the founding of the People's Republic have laid the bulk of contemporary ecological problems at the feet of the Qing and the Nationalists. In discussing the Keerqin Sandy Lands, for example, a recent publication explicitly blames the Qing for the most recent round of ecological devastation in the region. Charting the ebb and flow of local desert conditions over the past ten centuries, the publication contends that

by the beginning of the seventeenth century, Horqin [Keerqin] had thrived again, with tens and thousands of horses, camels, sheep, and cattle grazing and breeding on the pastures. But after the middle of the nineteenth century, the Qing Dynasty pursued a policy of encouraging people to reclaim wasteland. People were allowed to open up pastoral land and grow crops by paying taxes to the court. In 1907 alone, more than 800,000 hectares of grasslands were destroyed in the area of Horqin Right Wing Central Banner, while the Qing government received an income of 238,000 taels of silver. The destruction of forests and grassland made way for wind and sand which gradually encroached upon the denuded lands. Horqin was turned into the 800 li of deserts. (BMOF 1990: 22)

Similarly, Zhao and Xing (1984: 247) primarily implicate prior governments, basically absolving the People's Republic of culpability. It was under the Qing, they write, that "large tracts of sandy lands in the south-western part of the Ordos Plateau were ruthlessly cultivated, resulting in further devastation of grasslands and an extension of the shifting sand dunes. . . . Henceforth, coupled with accelerated human intervention, desertification has been critically intensified." They date the most critical desertification activities to the ninth through fifteenth centuries, but they also detail further expansion over a period of 300 years from the mid-Ming right up to 1949.

The Republican and Nationalist eras of government are likewise prominent targets of criticism. In a statement typical of the Maoist era, one reporter explained: "Before liberation the feudal ruling class, Kuomintang reactionaries, and imperialists plundered and destroyed the forests, turning the north and northwest of China and the greater part of the loess plateau into regions nearly bare of trees" (Soong 1972: 23).

Government statements issued throughout the Maoist era tended to reinforce the sense of a magical cutoff date around 1949. In a publication prepared in 1975, for example, the national Department of Desert Research summarized the official view:

Over the years before liberation in 1949, the people living in China's desert areas were oppressed and exploited. As their natural resources were wasted and plundered, they were forced to retreat before the advance of wind-driven sands. Since the founding of the People's Republic of China, they have embarked on the mass movement, "in agriculture, learn from Dazhai." In the spirit of self-reliance and hard struggle that typified Dazhai, the famed agricultural production brigade, . . . comprehensive measures were developed in a cooperative spirit, with scientific and technical personnel working closely with the farmers. As a result, a number of achievements were realized, the basis for sand control established, and considerable progress in animal husbandry and agriculture recorded. (DDR 1982: 4)

Other scholars argued from case studies in eastern Inner Mongolia that the sandy dunes had been subjected to reckless cultivation, overgrazing, and deforestation until 1949, when government-initiated sand control measures began (in the mid-1950s) to stabilize and restore vegetation with tree-shelter belts (Zhao 1990: 263–270; Chonghalakoushu and Jisizhengli 1986: 105).[7]

A propaganda piece appeared in the early 1960s about Nasihan (Manduhu and Nasendelger 1963). The article is noteworthy because it illustrates how the official discourse of the collective era attempted to remove problems of production to an earlier period, and because it draws explicit connections between a history of economic exploitation and desert control—a subject largely absent from official discourse in the reform era. The theme of the article is that "where the rich failed, the poor have succeeded" in bringing prosperity to the grasslands.[8] Another article, written by a celebrity novelist who visited one of the new communes of Wengniute banner, gushed with enthusiasm: "With the coming of the people's communes, these sandhills and wastelands are truly being turned into a land of milk and honey" (Lao 1961: 13). Such exaggerated claims are especially interesting when contrasted with the (similarly hyperbolic) praise heaped on contemporary privatization policies by scientists and local government authorities for introducing a new era of prosperity.

In placing the blame on former political regimes, the intelligentsia of the Maoist era felt free to draw explicit connections between resource exploitation and social exploitation. State-run newspaper articles consistently glorified the "liberation" from nature that followed the "liberation" from feudalism with titles such as "The Desert Surrenders"; "We Bend Nature to Our Will"; "How We Defeated Nature's Worst"; "Hard Work Conquers Nature"; "The United Will of the People Can Transform Nature" (see Murphey 1967: 319; Salter 1973).

Despite the rhetoric of good stewardship and mastery over nature, however, the Maoist era was not so kind to the national rangelands. After taking power, the new government did initiate some new programs and methods to control moving sand dunes. For example, it established and funded the Institute for Desert Research in Lanzhou to conduct experimentation and research on dune fixation techniques. And collective organization in the north motivated some aggressive experimentation in land rehabilitation. Yet, relative to other programs and concerns, land degradation in border regions did not receive all that much attention at the national level. As one scientist in Lanzhou (Dr. Ju Gebing, vice president of the Directorate for Environmental Protection) put it in his opening remarks to a United Nations Environment Programme (UNEP) seminar hosted by China in 1978: "For more than thirty years since the founding of the People's Republic, work has been done and some results achieved in the control of desertification. But still we would have to say that our work has just begun, that there lies before us an arduous and long-term task" (quoted in Walls 1982: 59).

The early commitments to control desert expansion in the 1950s lost out to other priorities in subsequent decades, as blueprints for the development of the national economy changed (*Renmin ribao* 1991; *China Daily* 1991). Through the long series of collective-era production campaigns, officials reminded local commune leaders repeatedly of the secondary value of protecting the rangeland in relation to other, more pressing objectives, such as increasing grain output and industrial production. During the collective era, national rangelands came under unprecedented demographic strain, yet the amount of money invested per unit of area in pasture improvement was less than one-seventieth the value of animal husbandry products per unit over the same time period (Watson, Findlay, and Du 1989: 226).

*Absolve the reform era.* The reformers who came after Mao added the Maoist government to the list of those culpable for the nation's ecological problems. While they consider the Mao years to be less neglectful than those of the Qing or the Nationalists, they nevertheless use them as a foil against which to prove the superiority of their own policies. The year 1978 has become a new magical cutoff date for the rhetoric of ecological responsibility. For example, Zhao Songqiao of the Institute of Geography (Chinese Academy of Sciences) in Beijing has written that

since the founding of the People's Republic of China, great efforts had been taken in the 1950s to combat this desertification process. . . . Then came the so-called Great Leap Forward and Cultural Revolution periods. . . . This led to a dramatic acceleration of the desertification process. . . . Since 1978, great efforts have been again undertaken to harness the Mu Us Sandy Land. . . . Thus, the desertification process is now getting checked, and the de-desertification process is asserting itself. (1990: 265–266)

Likewise, Zhu Zhenda of the Department of Desert Research has written that the present desertlike features across much of the northern landscape have been shaped chiefly over the last 50 to 100 years, but mostly in the last half-century. In the area of the Keerqin desert, he asserts that human-induced desertified land increased from 20 percent of the total area in the 1950s to 53.8 percent by the end of the 1970s (Zhu Zhenda 1990: 62, 65, 70). In Wengniute banner, grassland scientists have claimed that "reckless" (*wu jiezhi de*) land use intensified especially over the last 30 to 40 years (see Kou and Xue 1990).

Frequent praise for a massive afforestation program initiated in 1978 represents another case in point. This project—dubbed "China's Great Green Wall"—is described as the "top ecological undertaking in the world," a tremendous feat of engineering (Li 1990: preface). The official spin is that before 1949, "ruthless" deforestation led to widespread land erosion, but after the founding of New China, the people of the Sanbei Shelterbelt Region devoted themselves to afforestation, transforming the denuded mountains and harnessing the drift sand.[9] After twenty-eight years of experimentation, a concrete plan for the nationwide development of shelterbelts was put forward. Since then, officials claim, there has been marked improvement of the environment (even as degradation accelerates). Premier Li Peng cited the Northern Shelterbelt as evidence that China (in the post-reform era) has "vigorously promoted scientific and technological research on [the] environment." He praised the project as a "Great Wall against sandstorms" (China State Council 1994: 3).[10]

In contemporary China, the struggle to control the desert is often contextualized in just such a discourse of modernity, invoking the prowess of advanced scientific technologies to dispel the ancient threats of sand drift that menace more backward societies. The consistent political message conveyed to the public since the reform era has been that thanks to the technological harvest of the reform-era modernization process, we will finally subdue and control our northern deserts.

Glowing reports of technological developments surface repeatedly in the media, and they usually tap into both or either of two themes that play an important role in the official discourse: science and internationalism. These themes help to identify the reform era with "modernization" and "globalization." For example, one of the most favored recent technologies is aerial seeding. China began to experiment with the technique of broadcasting grass seeds from airplanes in 1979, and the successful aerial seeding of an arid region was reported in the press as a great breakthrough, made all the more impressive because stunned foreign experts had believed it impossible (*China Daily* 1988). Proponents of afforestation projects also hope to achieve greater public reverence by invoking an aura of scientism. A recent news report informed readers that "the composition of [the] shelterbelt forest system was based on countless laboratory experiments involving computer modeling and wind-tunnel tests. As a result, the shelterbelt forest was planted in a configuration designed to provide optimum protection for vegetation and the surrounding environment" (Jiang Wandi 1994: 18). Another new method to combat desertification that appears in the media is water-saving biotechnology. Spokesmen at the Soil and Water Conservation Institute under the Chinese Academy of Sciences have proudly reported the development of a chemical that can absorb and release large quantities of water that might be used to promote agriculture and afforestation in arid regions (Xinhua 2000e). Lately, media reports have announced government intentions to breed improved varieties of grass inside satellites that will be launched into space by the year 2003 (Xinhua 2000f).

The political importance of the highly publicized campaign for desert control is further illustrated by the number of bureaucratic agencies (at least sixteen) now charged with responsibility for carrying out the national antidesertification campaign. Indeed, the complexity of the bureaucratic structure seems designed more for the propaganda value of portraying a comprehensive team effort than it does for coordinating effective solutions.[11]

In summary, past and present political factions in China have found great propaganda value in showing themselves to be at work in taming the desert and in appearing to be more effective at the task than their predecessors. Through public discourse, intellectuals within the current regime claim to have done much to ameliorate the inherited legacy of irresponsibility, yet they also concede that deterioration has not been arrested, mostly because of irrational land use among ignorant or backward herders and farmers who continue to resist modernization. While often critical of shortsighted land use policies from earlier periods, Chinese authorities primarily scapegoat local rural producers.

## Environment, Politics, and Cosmic Harmony

Why would Chinese officials be so concerned to deflect the blame for land degradation away from themselves—what makes their culpability so dangerous? Obviously, there is no single answer, but the long tradition and lasting influence of Chinese natural philosophy provides one path of explanation. Throughout the history of imperial China, the natural environment was conceived primarily in the context of political harmony. For millennia, government authorities based their legitimacy on the notion of a "mandate from heaven." The Emperor, as Son of Heaven, was responsible for maintaining harmony between Heaven and Earth. Evidence of proper governance was manifest by harmony in both the social and natural order. By the same token, natural disasters could be construed as evidence of disharmony—ordinary citizens associated them with incompetence among the ruling elite and perceived them as a sign of discontent on the part of Heaven (see Needham 1956: 359–363; Huffman 1986).

This natural philosophy found support in the traditional art of *feng shui*. *Feng shui* beliefs and practices permeated Chinese society, influencing people to be closely attentive to nature-related symbols and to the possibility of writing symbols into nature (Bruun 1995: 184). Grapard (1994) notes that believing in an ideal harmony between the structure of the world and themselves, humans were on the lookout for such signs in nature. The world was then conceived as a text to be decoded. . . . A corollary of these views was that humans might see themselves as the agents of cosmic change, so that whenever disastrous events occurred in the natural world, they embarked on protracted rites of penitence to pacify the moral reactivity of nature, and whenever auspicious events occurred, they performed rituals of gratitude. . . . There could not be a single natural phenomenon without its corresponding cultural echo. (380)

Given this natural philosophy and its pervasive influence, it is not surprising that dynastic rule itself was sometimes the victim of natural disaster in China. Widespread devastation and social turmoil resulting from floods, earthquakes, and famine have historically been major contributing factors to the collapse of imperial authority. For example, the worst drought in the past 500 years hit northern China in the waning years of Ming authority (1634–1643). It contributed to a veritable army of refugees heading north that precipitated social unrest and the eventual fall of the dynasty (Reardon-Anderson 1995: 55–56). These ancient associations have not disappeared with the socialist state. People reacted nervously, for example, when a terrible earthquake hit near Beijing in 1976 (not long before the death of Mao) that claimed 665,000 lives. A popular slogan, charged with national political significance, circulated widely after the event: "Criticize Deng, resist quakes, and recover from disasters" (Huffman 1986: 75; *Renmin ribao* 1976).

The doctrine of cosmic reaction to political governance was by no means unique to China, but nowhere else was it "enshrined as a central part of a philosophical and moral system" (Murphey 1967: 314). Since 1949, however, the Communist Party has tried to institute a radical departure from the traditional views linking nature and governance. In the words of Murphey: "Nature is no longer to be accepted but must be "defied" and "conquered." . . . Nature is explicitly seen as an enemy, against which man must fight an unending war" (319). To some extent, the new rhetoric of conquest has only intensified the political and potential symbolic importance of nature and the environment for Chinese authorities. Once the metaphorical gauntlet was thrown down, the state could hardly appear to be losing control.

Even into the reform era, many environmental issues have taken on highly symbolic political meaning. National leaders, for example, have been hesitant to permit the growth of green activist organizations for fear that it will serve as a launching pad for political opposition (Lam 1993;

Associated Press 1994). Also, in the debates over development programs in recent years, environmental issues have become symbolic means to question the political and moral legitimacy of factions within the government, if not the entire Communist Party. National debates over wildlife conservation (see Schaller 1993; Carpenter 1989) and the Yangzi River Three Gorges Project (see Dai 1994; Sullivan 1995) are two of the most conspicuous examples. The longstanding public expectation of responsible domination over nature still influences the political process.

China's official discourse about deserts and rangeland policy, therefore, has been neither casual nor unbiased. It affects not only how scholars and officials gauge the scope and severity of degradation, but also how they direct public interpretation of the causes and the culprits and the symbolic significance of desert land. This language perpetuates an important representation that the state has created about itself and the effectiveness and benevolence of its policies in minority areas. The reality of this discourse, and its power to construct knowledge on environmental issues, too often lies hidden behind the authority of scientific pronouncements.

# THREE

# The Ambiguities of Land Degradation

The widely publicized "facts" of grassland degradation, upon closer examination, are not so straightforward as they first appear. Quite the contrary, for a number of reasons they are deeply ambiguous, prone to manipulation, and instrumental in concealing the routine exercise of social power. Indeed, I consider the conventional accounts of official discourse to be prominent obstacles that get in the way of potential insight and deeper understanding of serious environmental problems.

It is my intention to help redirect and expand popular and scientific understanding of land degradation processes—in China and elsewhere— by drawing attention to neglected social and cultural considerations that condition the way environmental changes are publicly defined, structured, and evaluated. I do not dispute the general claim that Chinese grasslands are in a state of degradation and acute reconstruction. I do, however, dispute several prominent assumptions of the official and scholarly discourse in China: first, that knowledge of land degradation is objectively constructed; second, that the current government can justifiably deflect culpability for the crisis by blaming others; and third, that current rangeland enclosure policies are saving the grasslands and establishing a base of security and progress for local residents. I also dispute some prominent assumptions in Western scientific literature about land degradation as it occurs in developing societies: first, that cultural factors are relatively insignificant compared with political-economic factors; second, that privatization offers a tidy solution to land management problems; and third, that the structure of Western science does not itself limit our ability to understand resource management problems in unfamiliar settings. All of these popular misrepresentations combine to reinforce the naive but convenient view that ignorant herders and farmers are the primary agents responsible for destroying the grasslands of Inner Mongolia, and they work to conceal the many legitimate reasons for redefining the environmental crisis as a profound problem of state management that has only been magnified in recent years by the penetration of global capital and influence.

I will begin to develop these arguments over the next two chapters, but it will take the rest of the book to conclude them. The immediate point, however, is that land degradation (and desertification)—like any other environmental concern—is inherently a socially constructed and subjectively defined phenomenon that cannot easily be squared with the language of scientific fact. There are fundamental problems involved in trying to define the terms, assess the damage, pinpoint the blame, and propose practical solutions.

## Problems of Definition

"Land degradation" usually implies a loss of intrinsic soil qualities necessary to sustain an economically viable agriculture, whereas "desertification" generally refers to "irreversible" human impact in arid, semiarid, and dry subhumid environments. Critical changes in soil structure and texture typically result from a combination of physical, chemical, biological, and socioeconomic factors that may be initiated by a variety of land use pressures including deforestation, grazing, farming, and industrial activities. In everyday language, it seems easy enough to grasp the idea of degradation, but nailing down the specifics in rigorous language presents many conceptual difficulties.

First of all, the word "degradation" conveys no precise meaning in any scientific or economic sense. The same is true for the Chinese term, *tuihua*, which simply denotes a movement backward.[1] To call something "degraded" (in English or Chinese) is to assign a negative moral value to a morally neutral physical process. That assignment necessarily reflects the political, economic, and cultural interests of a particular social group.

The term "desertification" is even more controversial and difficult to define. More than 100 definitions appear in academic literature since 1980. The term's elusiveness springs primarily from the difficulty of incorporating into scientific language the distinctions between drought-induced changes that sudden precipitation may reverse, and more permanent changes associated with prolonged human interference (Thomas and Middleton 1994). Noting the problems, Swift (1996: 73) has argued that the concept "has less to do with science than with the competing claims of different political and bureaucratic constituencies." It might be preferable to avoid the term altogether, but Chinese officials and scholars use it (as well as comparable terms such as "sandification") routinely in reference to North China and Inner Mongolia.

There simply is not, nor can there be, any universal consensus on a definition for land degradation (or for desertification and other derivative terms). Still, an international community of concerned scientists has managed to agree on two basic principles: there must be a significant and lasting decline in biological productivity, and it must result from both natural forces and human activity (UNEP 1993: 133). Yet even these modest principles are extremely vague and subject to intense debate. How does one quantify and measure such imprecise terms as "significant," "lasting," "decline," and "productivity"? And how does one distinguish "natural" forces from "human" activity? There are no clear and absolute standards to apply.

Consider, for example, the ambiguous notion of "significant and lasting decline." Any attempt to determine whether soil changes are significant or insignificant, temporary or long-term, leads inevitably to the tension between "sensitivity" and "resiliency." Sensitivity refers to the susceptibility of a given land system to unfavorable physical transformations (however those may be defined), while resiliency refers to the capacity of a given land system to absorb the effects of interference and yet remain productive (Blaikie and Brookfield 1987: 10). The natural resiliency of any land system has limits and, over time, requires supportive land management techniques such as crop-fallow rotations, fertilization, irrigation, and so on. A land system characterized by high sensitivity and high resiliency degrades easily but responds well to proper land management. A land system of high sensitivity and low resiliency degrades easily but does not respond well to land management. In an arid environment such as Inner Mongolia, sensitivity is usually considered high and resiliency low.

The conventional perspective holds that dryland ecosystems are fragile and, once disturbed, require "decades, or even centuries, to regain productivity" (NRC 1990: v). Yet skeptics argue that dryland ecosystems are actually far more resilient than presumed (Johnson 1979). Some ecologists even maintain that dryland ecosystems are at least as stable as humid ones, and maybe more so. Others simply object to the way the notion of fragility passes so easily as established scientific doctrine (Mortimore 1988: 62). There are comparable debates over how to define and measure many other key terms.[2] These gentle controversies in Western scientific literature are relevant to Inner Mongolia because they underscore both the complexity and the subjectivity with which pastoral Mongols and urban Han disagree over what kind of landscape deserves to be considered "productive," and what kind deserves to be considered "wasteland."

Conceptual ambiguities also cloud any attempt to assert the real cause of degradation. Who, for example, are the most critical actors—the local residents who actually work the land, or individuals and institutions in distant centers of power that structure the political economy in which local decisions are made? Since land degradation is always situated in a particular locale, it seems reasonable enough at first glance to look for causal explanations in the behavior of local inhabitants. But there is no legitimate reason to limit the search for culpability to the geographical boundaries in which degradation occurs. On the contrary, there are many good reasons to connect local production with broader systems of human use. The more local land users are drawn into production for sale in national and global markets, the more their land use decisions are vulnerable to external influence. China's agricultural producers, both cultivators and herders, have borne a heavy share of the burden of national development, and their burden has been displaced onto the soil.

## Realities of the Chinese Political Economy

The rangeland continues to be neglected in the post-Mao era. Although the current regime may have thrown more money at the problem of land degradation than did previous governments, it does not necessarily follow that they have intensified efforts on the ground to control desert expansion in meaningful ways. Indeed, many foreign and domestic observers attribute the deepening environmental and ecological crises all over China to the decentralization of post-Mao economic reforms.

William Hinton (1990: 21) has been particularly outspoken in linking local resource mismanagement to national economic mismanagement: "Thus began a wholesale attack on an already much abused and enervated environment, on mountain slopes, on trees, on water resources, on grasslands, on fishing grounds, on wildlife, on minerals underground, on anything that could be cut down, plowed up, pumped over, dug out, shot dead, or carried away." Vaclav Smil (1993: 195) has asserted: "After two decades of suffering and material deprivation, the Chinese countryside of the 1980s was in no mood to join en masse any movement aimed fundamentally at conservation and efficiency. Not surprisingly, accelerated consumption, accompanied by waste, pollution, and degradation, was the tenor of the decade." Richard Smith (1995: 34), who focused more upon the national elite, has drawn a similar conclusion: "China's entrepreneurs and their multinational corporate partners, each pursuing their own self-interest, each acting without a coherent plan or any assessment of economic, social, or environmental impact of those decisions are, in the aggregate, driving China's economy down the road to environmental barbarism."

Many other scholars have pinpointed rather specific circumstances within the Chinese political economy since 1980 that contribute to ongoing land use problems. These circumstances may be grouped around the following themes: low investment and administrative neglect, insecure and limited tenure, unfavorable prices, insufficient labor supply and alienation, miscellaneous inefficiencies, the irrelevance of law, and an adverse national development strategy.

## LOW INVESTMENT AND ADMINISTRATIVE NEGLECT

The reform government has invested much less in animal husbandry than it has in crop cultivation and forestry. According to Zhou Li (Institute of Agricultural Economics, CASS), "only a pitifully small portion of scarce [agricultural development] resources has gone into improving livestock production" (1990: 48). In Inner Mongolia, he argues, investment in livestock production constitutes only 10.8 percent of total agricultural investment, though livestock accounts for 26 percent of total agricultural output (see also Liu Guizhen 1993: 2). Likewise, relatively few research or administrative resources have been devoted to sheep and wool industry problems (Longworth and Williamson 1993: 2).

Out of the meager funds invested in animal husbandry, the government earmarks only a small portion for pasture improvement. Zhou reports that state subsidies for pasture rehabilitation have never been included in any five-year plan at the central government level (Zhou Li 1990: 47). Without such support, the pastoral provinces and autonomous regions do not have adequate resources to support grassland development. Zhou also reports a failure to invest in technologies that might exclusively benefit the grassland population, such as superior arid zone plants that might resist extreme temperatures and drought conditions. Likewise, insufficient planning at the national level has resulted in critical shortages of useful material supplies (ibid., 48). Liu Yuman (of the Institute for Rural Development, CASS) has specifically linked accelerating land degradation in pastoral areas to a lack of funding in macro-level management to control animal populations and pasture reconstruction.

He argues that this situation is not new to China, but has become "much more serious since the early 1980s" (1990: 97–98).

Even when the region does manage to attract large investment capital, local herders are likely to be bypassed in the development process. For example, the new "Grain Distribution and Marketing Project," involving a total investment of about U.S.$1 billion, has recently begun in the Northeast. The World Bank has extended a twenty-year loan in the amount of U.S.$325 million to support the government's grain sector reform program, which intends to achieve greater efficiency in grain production and marketing on a national scale. Specifically, the project hopes to introduce for the first time in China a large-scale, fully integrated, bulk grain logistical system in three major production and consumption corridors. The major production corridor is the Northeast, and Chifeng City is one of many depots along the transportation route to be upgraded. Apparently, surplus feed grain from Heilongjiang, Jilin, and Liaoning provinces will be shipped by rail to the Yangzi River valley and the Southwest (especially Guangxi province) to be consumed there by the livestock of predominantly Han farmers who stand to increase their production for market (Song 1994). It seems that there are no plans to channel any of the grain to neighboring areas of Inner Mongolia, where chronically undernourished livestock might be fattened and shipped out (with value added) to achieve the same national goals of market development and reductions in transport cost for the outflow of grain.

Nor does the problem of wind and soil erosion in desert-susceptible areas enjoy a high investment priority. Although I have not been able to locate comprehensive government statistics on past expenditures for disaster mitigation, scientists within the Shenyang Institute of Applied Ecology in Nasihan assert that desert monitoring and control efforts have historically received much less money than other national hazards, such as flood, fire, and earthquake, which generally affect more populated regions (that tend to be inhabited by ethnic Han). This arrangement will apparently continue well into the future.[3] The director of the Lanzhou Institute of Desert Research remarked that Beijing's input on desert control was so limited that "the local governments and people themselves had to raise most of the money needed to solve the problem." He further complained that the four to five million yuan that the institute annually receives from Beijing is barely enough even to pay the salaries of its 300 staff members (Xia Xuncheng, quoted in the *South China Morning Post* 1994).

It is not surprising, therefore, that residents who live in areas threatened by desert encroachment find the constant government rhetoric of self-congratulation ridiculous. In Nasihan, for example, the grassland ecosystem research station displays a plaque presented by the People's Government of Wengniute banner, which praises the Han scientists with a couplet: "Heart blood has become sweet dew, the desert has become an oasis" (*xin xue hua gan lu, sha mo bian lu zhou*). Village residents scoff at such platitudes. One old herder told me that if he had written the verse, it would state the reverse: "An oasis has become a desert under the management of the research station." Some residents use the colloquial phrase *bimen zaoche* (load the cart behind closed doors) to refer to the way in which local cadres and scientific tourists invent official statistics to suit their own purposes, building a grand public facade to hide unpleasant truths.

## INSECURE AND LIMITED TENURE

A recurrent problem with national-level policy that has had significant consequences for local land use practices has been the insecure and limited terms of household ownership for both livestock and land. In some areas of pastoral China, over the span of just one generation personal property rights have changed some nine times since the Communists took control in 1949. Variations in the exercise of livestock ownership have especially fluctuated through several stages, moving toward both collectivization and then decollectivization.[4] With regard to land tenure relations, substantive changes have occurred at least four separate times since 1949, not including all the instances since decollectivization in which the central government announced relatively minor alterations to the contractual duration of private land use rights. When independent rural households first began to assume responsibility contracts for land use, terms of tenure were indefinite. They were later set to three years, then stretched to fifteen years with rights of inheritance. Today, they are set at fifty years, but the crisis of confidence in land tenure understandably remains acute for both herders and farmers across the nation.

For both Han and Mongol cultivators of Inner Mongolia, insecure land tenure contributed to the neglect of soil conservation practices through the early 1980s. Given the state of flux, land managers learned to fix their sights on short-term profits rather than orient their labor or their fields toward long-term sustainable yields. On the topic of composting, for example, Pasternak and Salaff (1993: 110) report the bitter words of one informant: "There are good reasons not to take the trouble. It may be good for the land but, given recent history, I could end up working for someone else's benefit. They have already reallocated land twice. What's to stop them from doing it again? Each time they do that, the labor and effort you have invested collecting and processing compost becomes a gift to someone else!" Insecure tenure in the early years also interfered with the ability of diligent farmers to manage a plot of land efficiently even when they wanted to, since the crop rotation patterns of previous land managers were usually unknown. This put both the farmer and the land at risk, as production levels would decline by 50–70 percent if the wrong crop was planted at the wrong time (ibid.: 60).

Grassland herders have faced comparable institutional constraints. Longworth and Williamson (1993: 313–321) argued that the early prohibition from selling or even leasing rangeland contracts to others encouraged herders to "mine" their pastures for short-term benefits. Since other factors of production such as livestock and machinery could be freely traded, they argued that households would rationally underinvest in improvements to pasture land through such means as fencing, digging wells, or seeding. And even though herders in some areas are now able to lease their land use rights, tenancy arrangements are not likely to reverse the emphasis on short-term gains. Ho (1998) has argued, based on a case study in the Ningxia grasslands, that the mass confusion of ownership rights that began with unfamiliar institutional arrangements during the collective era has only deepened since 1980, so that overgrazing is systematically encouraged rather than managed.

UNFAVORABLE PRICES

Real incomes in some pastoral areas of Inner Mongolia have not improved since about 1983, whereas income levels have increased significantly in most agricultural regions of China since the early 1980s (Longworth and Williamson 1993: 101). In part, this is because comparative price advantages have moved in favor of crop production relative to animal husbandry. For example, the price of grain tripled in the 1980s relative to the price in the 1950s, but the price of live sheep increased by only a factor of 2.6 (Zhou Li 1990: 50). Herders will not voluntarily restrain herd growth under such unfavorable terms of trade. When times are bad, the majority still seek to minimize risk in the traditional manner of accumulating more animals.

Other market forces have also directly encouraged opportunistic stocking strategies. One glaring problem is that procurement prices do not vary appropriately between commodities of differing quality. Better-quality beef does not command premium market prices, for example, so that herders have no incentive to kill off young cattle rather than carry them through difficult winter months on expensive and poor quality feed. The inevitable loss of weight amounts to great wastage of pasture resources consumed in the fattening process (Hinton 1990: 89). State pricing mechanisms for mutton and wool also greatly affect flock structures and production pressures. Throughout the early 1980s, market incentives promoted an increase in the output of mutton and a decrease in the output of wool, which led to bitter competition among various enterprises in the purchase of wool. As a result, central authorities eventually lost control over regulation of the market. In turn, they lost the ability to effectively moderate pastoral flock structure and growth rates (Watson, Findlay, and Du 1989: 234; Liu Yuman 1990: 98). Hinton believes that with a proper herd and flock structure, "half the number of cattle can produce twice as much meat . . . and a somewhat reduced flock of sheep can produce three times as much meat and 50 percent more wool than is now generally obtained" (1990: 86).

### INSUFFICIENT LABOR SUPPLY AND ALIENATION

Rural decollectivization has had profound implications for the supply of agricultural labor, which in turn affects land use decisions. On the one hand, land fragmentation has destroyed economies of scale, so that agropastoral mechanization is a financial impossibility for most households. As a result, production depends heavily upon the supply of labor at crucial moments in the agricultural season such as planting, harvesting, mowing, and shearing. There has been such an exodus of labor from many agricultural areas that the remaining labor force is often unable to cope with the crops and livestock. On the other hand, the requirement for abundant cheap labor is not met. The high cost of wage labor prohibits many families from hiring the necessary manpower to maximize either agricultural production or conservation over the long term. (Short-term work parties and household labor swaps that avoid the need for cash are still common practices.) Furthermore, when the millions of migrant workers reach their urban destinations, they cannot obtain residency rights, leaving their home communities to bear the cost of educating their young and caring for their elderly.

Closely related to bottlenecks in labor supply is the spirit of isolation brought about by decollectivization and land fragmentation. Informants from Inner Mongolia confirmed to Pasternak and Salaff that few farmers or herders are eager to cooperate with others. "Some of us do cooperate on a short-term basis . . . but we rarely continue such short-term cooperation or extend it to other activities" (1993: 71). Humphrey and Sneath (1999: 169) also quote informants from Inner Mongolia who assert that people "are not so helpful and friendly with others as they were in the collective period." Howard (1988) has also argued that the decline in rural cooperation contributes to inefficient land use in other significant ways: a decline in water conservancy infrastructure (56); idle capital scattered among independent household accounts (107); and untapped technical and entrepreneurial ingenuity that remains buried under the constraints of underfinanced production efforts (123).

## MISCELLANEOUS INEFFICIENCIES

Grassland residents face severe constraints in their production capacities because of poor infrastructure. Residents generally lack local canning, dehydrating, and freezing operations, leading to terrible waste in food spoilage and lost animal by-products (Howard 1988: 123). Poor transportation infrastructure contributes to the problem of spoilage and, more significantly, poses serious limitations for economic diversification options out of animal husbandry.

Other widespread rural problems include the lack of affordable sources of energy from two to four months every year, which perpetuates the use of organic waste as household fuel. Residents scour the countryside to collect dried cattle dung to burn in their hearth fires. Related to this problematic drain on soil nutrients is the increasing preference for synthetic fertilizers, so that composting and green manuring practices have basically subsided (Smil 1987: 221).

## IRRELEVANCE OF LAW

The lack of an effective legal framework to penalize abusive land management has been cited by numerous authors. Only in 1983 was the first Grassland Law issued to prohibit unsanctioned cultivation and to encourage farmers to convert cultivated lands back into pasture. Yet land management enforcement remains lax, as farmers typically plant grass in the spring to qualify for tax breaks and other government incentives, only to break the sod again after rewards have been secured (Ma 1984: 158–159). Regional and local officials play similar games by rewarding individuals and entire communities who prosper through their disregard of environmental regulations. Despite legal restrictions on goat husbandry, for example, large producers of cashmere attract the construction of government-supported factories, and they may even become a source of high-interest credit to the state (Humphrey and Sneath 1999: 104–105).

Since 1983, the IMAR government has adopted many regulations designed to regulate land use more strictly in the grasslands. The most important was the National Rangeland Law promulgated by the Standing Committee of the Sixth National People's Congress in 1985. This law itself fails to provide specific details about a number of critical management issues. According to Longworth and Williamson, "it was expected that the various provincial, prefectural, and county-level administrations would develop increasingly precise interpretations and guidelines for action" (1993: 83).

Despite the emergence of national laws, an essential problem remains one of enforcing compliance. Longworth and Williamson accurately invoke the term "policy mirage" to depict a significant problem familiar to anyone with experience in the Chinese countryside:

At Central government levels certain policies are in place and provincial, prefectural, county and even township officials will describe, often in considerable detail, how the policy is working. However, at the village and household level, the policy does not exist. In fact, the real situation at the "grass roots" may be almost exactly the opposite to that "seen" from higher levels in the administration. The higher-level officials are seeing a policy mirage. (ibid.: 322)

Understandably, many environmentalists cry out for more coercive enforcement of management regulations. Though the need for this "get tough" approach seems reasonable enough, I believe the emphasis is seriously misplaced. Without adequate provisions for the legal protection of individual rights, enforcement measures are likely to punish the wrong people. My field experiences lead me to suspect that any future campaign to penalize abusive resource managers would simply become a vehicle to exploit further the poorest herders who already suffer the injustices of unfair policy interpretation and implementation by local officials. Lack of policy enforcement is less of a problem than institutional lack of understanding about social realities in local communities.

NATIONAL DEVELOPMENT STRATEGY

It would be quite reasonable to interpret all of the production problems discussed above as mere manifestations of a deeper underlying reality: the grasslands have been a low priority in the Chinese national development strategy. Chinese officials clearly admit as much in their current high-profile campaign to "develop the West." Launched in January 2000, the initiative has provided a forum for the top leadership both to praise Deng Xiaoping as the visionary architect of a coordinated national strategy and to promise a sizable and frustrated population that a more equal distribution of wealth is just around the corner.

In the early years of decollectivization, Deng offered many colorful metaphors to legitimize the end of egalitarian ideology and the creation of a new economic structure that would allow some to "get rich first." With vivid trickle-down imagery, he compared economic reforms to a rising tide that would eventually lift all boats to new levels of prosperity. In the meantime, income inequalities among national macroregions, as well as within counties and small communities, would have to grow larger. Since 1979, regions endowed with good transport links or proximity to urban markets have indeed enjoyed relatively faster growth. At the same time, the fiscal power of the central government to redistribute national investment funds has diminished, further accelerating uneven regional development (Knight and Song 1993: 202). Eventually, central government concern for regional inequality lost out to theories of regional comparative advantage, which favored heavy national investment along the eastern seaboard. The people of central, western, and northern China were encouraged to sacrifice their own well-being for the greater national interest. Throughout the 1980s, this strategy resulted in disproportionate benefit to ethnic Han Chinese over the national minorities whose population is concentrated throughout the periphery (Mackerras 1994: 227–228).

Now, after two decades, Chinese leaders have announced that the time is ripe for the eastern region to return the favor and help those who live in the far western regions of the country—places such as Ningxia, Gansu, Shaanxi, Sichuan, Guizhou, Yunnan, Qinghai, Xinjiang, and Tibet. Premier Zhu Rongji (2000) has identified the immediate objectives to be the construction of infrastructure (especially highways, railways, airports, oil pipelines, electric power grids, and telecommunications) and environmental conservation (especially of fresh water and natural forests). Given China's current surplus of food grain, the leadership considers it a good time to oversee the restoration of woodlands and pasture in the grasslands. It remains to be seen, of course, whether officials will make good on their latest promises, whether showcase infrastructure projects will spur lasting economic growth, and whether the large regional disparities of wealth will be reduced in a reasonable time frame. Economists within the Chinese Academy of Social Sciences have predicted substantial improvements to the area by the year 2030 (Xinhua 2000g).

Skeptics have already pointed out that the campaign appears designed to promote national solidarity but not to shift significant privileges away from favored eastern cities in any meaningful way (see Gilley 2000). For example, central government bank lending is scheduled to increase by only 5 percent over the next five years, with a one-time grant for tourism, education, and high-tech industries. And much of the infrastructural development seems geared toward the extraction of resources rather than the creation of processing facilities or increased consumption within the region itself. In the words of Li Peng, "the state should implement the principle of sending the western gas and electricity to the east, for this will improve our country's energy structure by a large margin" (Li 2000). But if the people living in the west really hope to improve their standard of living, surely they will need significant policy changes at least as much as they need infusions of money. And even if government rhetoric does lead to action, the herders and farmers of Inner Mongolia and other excluded regions (of "central" China) will apparently have to keep waiting for their turn at the public trough. Until that time, since they are neither east nor west, they remain, in the words of a bitter Mandarin pun, literally "nothing" (*mei dongxi*).[5]

The central government has not been willing to make the expensive and large-scale capital investments that would be required to develop the arid rangelands along the northern frontier into the kind of productive regions that officials claim to desire. The land and the people who are most directly exposed to the threat of desert expansion remain economically and politically marginal in the scheme of central government concerns. Future policy initiatives may yet improve the long-term prospects of some grassland areas, but there are many reasons to remain skeptical.

Despite prevalent Han stereotypes that traditional Mongol culture devalues and destroys the land base, there is ample evidence to support a contrary view: the Han central authorities have systematically undervalued the social worth of national rangelands. On the one hand, officials sound the alarm of degradation but, on the other, neglect and exploit the land and its inhabitants, thus perpetuating the acceleration of soil erosion. Given the realities of the Chinese political economy—especially the history of colonization, administrative neglect, and extraction of resources that has occurred over the last century—the popular discourse that attempts to blame local people exclusively for the environmental problems they now face must be recognized as disingenuous at best. By the same token, there are legitimate reasons to link both land degradation processes in Inner Mongolia and the national rangeland policies of the reform era with the broader international political economy.

### Linkages to Global Capital

Modernization and globalization have not always been synonymous in China. Mao tried vigorously to keep the modernist project separate from the influences of the capitalist world system. He set a course of economic development for the nation whose guiding principle was autarkic self-reliance. For roughly thirty years, China curtailed its participation in international trade with capitalist nations and tried to isolate itself from inegalitarian influences. The death of Mao in 1976 created an opportunity to reconsider these dogmatic positions. Deng Xiaoping became the champion of reform-minded officials who believed that further development required China to restructure its economy to benefit from the capital, technology, and models of development available from Western industrial countries. Only since the reformers initiated a new "Open Door Policy" has the modernist project of the Chinese Communist Party coincided with the forces of globalization.

The critical turning point came in December 1978, at the Third Plenary Session of the 11th Central Committee. The Party announced that China would adopt a number of major new economic measures to "consciously transform economic management, actively expand economic co-operation on terms of equality and mutual benefit with other countries . . . and strive to adopt the world's advanced technologies and equipment . . . to meet the needs of modernization" (CCP 1978: 11). As a first step, the plenary session adopted the view that China must focus its energies on agricultural advancement through privatization and the end of egalitarianism.

Just like that, a momentous historic shift had occurred that reached far beyond agricultural policies. The Party had expressed nothing new in the articulated desire for modernization. Rather, "what was new was the announcement that China would turn to the Western developed economies for advanced technology and assistance to achieve its objectives" (Ho and Huenemann 1984: 7). The Open Door Policy was perceived as the most direct path towards the long, unwavering goal of national self-reliance, as Premier Zhao Ziyang later explained in a prominent speech before the National People's Congress:

By linking our country with the world market, expanding foreign trade, importing advanced technology, utilizing foreign capital and entering into different forms of international economic and technological cooperation, we can use our strong points to make up for our weak points through international exchange on the basis of equality and mutual benefit. Far from impairing our capacity for self-reliant action, this will only serve to enhance it. (Zhao Ziyang 1981: 23)

Connecting China to the world market suddenly became an imperative. The Open Door Policy, however, involved much more than just trade liberalization. It was an outward-looking policy that envisioned technological advances purchased with the profits of expanding exports (Ho and Huenemann 1984: 21). In the mind of the reformers, such new global linkages were considered to be the key instrument of modernization. Through the early 1980s, Deng implemented a whole series of agricultural, industrial, commercial, and financial sector reforms that would relax centralized government planning authority and allow a more direct regulative role for international market forces.

EXPANDED TRADE AND COMMERCE

There are a number of reasons why contemporary grassland policies and the physical and social transformations that they have promoted should be linked to the operations of global capital. The most obvious is that they were explicitly intended to expand agricultural productivity for trade on domestic and international markets.

Consistent with the intentions of reform, rural decollectivization has basically been synonymous with increasing household commercialization. This has been true in animal husbandry as well as in farming. For example, in the period from 1978 to 1986, the volume of commodities purchased by Inner Mongolian herders reportedly increased by 20 percent yearly, while average household incomes from sideline production also increased by about 20 percent (Xinhua 1987). Once the privatization of pastoral resources was set in motion, the reform government in 1985 began to introduce more flexible procurement policies that allowed independent households to sell livestock and animal products to traders from regional metropoles. Though the official planned purchase system went through several permutations since then, by 1991 herders were free in principle to market their produce wherever they preferred.

Household commercialization of pastoral produce has expanded both domestically and in international circles since the early 1980s. As urban wages and living standards increased through the early 1980s, domestic consumer demand for meat, wool, and cashmere also increased. Zhou Li (1990: 44) reports that between 1979 and 1984, Beijing consumption of beef increased by a factor of 2.5, while the market for mutton doubled. Expanded output from pastoral provinces accounted for at least 90 percent of that growth. Likewise, urban consumers began to discriminate more carefully in the clothing and textiles they purchased, selecting more and better qualities of wool. Again, domestic production expanded so that the output of fine and improved grades of wool in 1991 represented an increase of more than 56 percent over that of 1981 (Longworth and Williamson 1993: 60, 329).

Chinese herders were also encouraged to expand production for sale on the world market. Early on, the national export strategy emphasized textiles and light manufacturing industries to finance imports and to generate hard currency for reinvestment in other sectors. This strategy relies heavily upon sheep and goat husbandry for a steady supply of wool and cashmere, as well as for many secondary animal by-products. Through the 1980s, the domestic woolen textile manufacturing capacity expanded rapidly, and China emerged as a major world player in the industry, both as a buyer of raw wool and as a seller of finished products (Longworth and Williamson 1993: 328).

Nasihan residents have participated in the expansion of pastoral produce for sale in domestic and foreign markets. For example, the Chifeng Government Trading Company operates a cattle grain-feed station in the northern suburb of Anqinggou. It is the nearest and most common marketing outlet for local herders, though a large percentage of cattle are also sold to independent traders from Shenyang, Chifeng, or Beijing who purchase directly in the villages. The Anqinggou stockyard fattens malnourished local cattle on corn and other grains for at least two months before shipping them live by rail to Hong Kong, where they are finally sold on the world market. Nearly 2,000 head of cattle are processed for export from this single feed station every year. Nasihan township alone supplies them with approximately 300 head per year (Li 1993).

EXPANDED INFLUENCE OF WESTERN INSTITUTIONS

A second reason to link grassland circumstances with global capital is that international development organizations have begun to play significant financial and advisory roles since 1979. Within Wengniute banner alone, two high-profile animal husbandry development projects—one funded by the United Nations Development Program from 1979 to 1983, the other, by the International Fund for Agricultural Development from 1981 to 1988—provided funds for a wide range of investments in pasture and forage production that specifically promoted wide-scale fencing. Regardless of how successful these projects may or may not have been in achieving their declared goals of modernizing local livestock production, they clearly enjoyed considerable leverage to promote the doctrines of carrying capacity and enclosure just as new animal husbandry policies were being drafted and implemented throughout the grasslands (see Chapter 6).

Parallel to this exposure at a regional level, at a national level China began to engage the administrative bodies of the world economic order during the same years. China and the United States established mutual recognition and full diplomatic relations on January 1, 1979, in tandem with the new economic program. China had become a member of the World Bank and the International Monetary Fund by 1980 and became a party to GATT's Multifibre Arrangement in 1984, requesting full participation by 1986. Participation in these organizations gave China new access to both world capital and foreign opinion, but it also committed its leaders to a series of international obligations that would make export-oriented economic reforms more difficult to reverse (Jacobson and Oksenberg 1990: 85). Through such new channels, international experts could engage Chinese officials in extensive dialogue about agricultural policy, thereby influencing the direction of resource management on specific issues (ibid.: 116).

For example, by mid-1983, the World Bank was primed to expand development loans to China. But first it wanted to analyze long-term development issues and options in the light of relevant experience from other nations. A subsequent report on the China livestock sector was eventually published in 1987, which expressed concern about increasing the amount of animal protein in the national diet: "Many of China's grasslands are overgrazed and probably cannot provide additional meat and wool in the short to medium term without further degradation. An urgent requirement is therefore to match livestock numbers with the land's carrying capacity, through herd reduction and improved range management" (World Bank 1987: 45). International loans became a vehicle to impose on Chinese grasslands Western blueprints for modernization through land enclosure. Through the power of the purse, foreign analysts have been able to influence the direction of China's rangeland development and privatization efforts.

Regardless of how successful or direct the influence of foreign actors may have been in accelerating Chinese momentum for specific grassland policies, the essential point remains: in order to gain access to foreign technologies and funds, Chinese leaders promoted new policies that were intended to expand commercial output on domestic and world markets, and permitted international development and trade organizations to exert significant influence over the direction of change.

## GEOPOLITICAL CONSIDERATIONS

In addition to the need for international capital and technology, Deng and the reformers were concerned to expand pastoral production for geopolitical considerations. The Open Door strategy brought China into greater contact with the international community. It simultaneously forced Chinese leaders to exercise greater caution with regard to the impoverished ethnic minority peoples who populate the expansive national frontiers. Specific to IMAR, Chinese leaders have been concerned about public protest demonstrations, independence movements, and calls for reunification with neighboring Mongolia (see, for example, Asia Watch 1992). Other obvious geopolitical considerations include Western focus on a variety of human rights issues and the promotion of international tourism among Chinese ethnic groups.

To appease the minority populations who were still traumatized by the violence of the Cultural Revolution, the central government made several conciliatory gestures. The death of Mao in 1976 officially ended the Cultural Revolution era and returned Chinese minority policies to more charitable dispositions characteristic of the early 1950s. A revised national constitution promulgated in 1978 conferred many privileges upon ethnic minorities: mosques were reopened, university examinations were offered for the first time in local languages, and non-Han were appointed to new showcase positions of regional authority (Connor 1984: 426–428). The connections between these gestures and the national defense were explicitly drawn when Chairman Hua Guofeng issued the following statements on the eve of the implementation of reform (1978: 34):

As to old and backward customs and habits, it is up to the people of the minority nationalities concerned to reform them step by step according to their own will . . . to give sincere and active help to the minority nationalities to develop their economy and culture is a major task in our nationality work, in building up our border regions and in consolidating our national defence.

Since the reform era, minority peoples have been allowed, even encouraged, to reassert many expressions of tradition in terms of clothing, religious festivals, and language, but assimilation remains the ultimate and unveiled goal of the party (Connor 1984: 428–430). The resurrection and conspicuous display of such "safe" manifestations of ethnic identity by both the state and local actors specifically to promote the industry of international tourism has been well documented (see Wu 1990; Gladney 1991; Oakes 1993).

The reformers have made a number of symbolic concessions specifically to the Mongol population. For example, in October of 1979, the Nationalities Committee of the National People's Congress reconvened for the first time since the Cultural Revolution, and Ulanfu, the formerly purged Mongol leader of IMAR, delivered a public address and was even permitted to make a strong case for meaningful autonomy (Connor 1984: 427). Then, on May 12, 1980, the Party rehabilitated the memory of Chinggis Khan himself. The *People's Daily* praised him as a "leader of Chinese and foreign peoples, an outstanding military strategist and statesman" (ibid.: 466–467).

Favorable territorial adjustments were also made to compensate for earlier losses. In 1969, authorities reduced IMAR boundaries by nearly one-third in a transparent effort to gerrymander Mongol demographic strength. Large regions of the eastern and western extremities were lost to neighboring provinces, and Wengniute banner itself temporarily fell under Liaoning jurisdiction. The partition created a situation where more Chinese Mongols lived outside IMAR than lived inside. But the earlier boundaries were abruptly restored in 1979 (Connor 1984: 323).

National attention to the economic plight of herding communities was a further manifestation of minority appeasement in consideration of security concerns. In the reform era, Chinese authorities intended to secure national borders by raising the economic level of grassland inhabitants. After decades of relative neglect, the central government suddenly addressed the animal husbandry sector in the Sixth Five Year Plan for Economic and Social Development, 1981–85 (Brown and Longworth 1992: 1664). The plan makes the following relevant commitments:

Individual households are encouraged to raise livestock and become specialized in this line. Support, help and advice are to be given to the various joint endeavors of crops and livestock farming, and integrated complexes of animal husbandry, industry and commerce will be organized according to the principles of voluntary participation and mutual benefit. . . . We will boost the production of forage and grasslands in the pastoral regions, expanding the man-made grasslands in 1985 to 100 million mu as against 32 million mu in 1980. In the south, grass hills and slopes will be properly used to raise cattle and sheep, on the loess plateau in the Northwest, we will promote planting grass and cultivating grass and grain in rotation. In agricultural, semi-agricultural, and semi-pastoral areas, we will popularize silage and improve silage methods. (Chinese Communist Party 1981: 58)

The minority populations were thereby promised a share in the rising economic fortunes of rural producers. This was necessary because Open Door policies made their welfare more important to the central government.

Given the influential role of the international political economy and the specific interventions of international development organizations on land use policies and practices in Inner Mongolia through the post-reform era, the culpability for ongoing degradation processes in the region can hardly be limited to local herders and farmers. It is clear that the multiple conceptual ambiguities of land degradation are only compounded by more practical ambiguities of locating responsibility and initiating reasonable measures to ameliorate the situation. The complexities of land degradation, however, are only half exposed as yet. In addition to considerations of the broader political economy, it is also necessary to explore how contrasting cultural systems of meaning play a crucial role in the transformation of the land.

# FOUR

# The Land in Cultural Context

Cultural realities—including attitudes, values, preferences, perceptions, and identities—can be just as important in shaping land use decisions as the material realities of political economy. I have already suggested that long-standing popular beliefs in a cosmic linkage between environmental harmony and political legitimacy might help explain some particular aspects of desert discourse in modern China. This chapter proposes a far more direct role for cultural factors in the process of land transformation, demonstrating that people relate to the land not just as individuals but also as members of a group with entrenched dispositions. In China, symbolic systems of environmental meaning actively shape the way rangeland policies are both designed in Beijing and implemented in Mongolian communities on the grasslands.

Ideological biases emanating from within the international community of scientists and scholars also have their effect on land use decisions halfway around the world. Contrasting worldviews and expectations among different cultural communities—urban Han Chinese, ethnic Mongolian pastoralist, and detached Western intellectual—thus constitute an important dimension of environmental controversy in Inner Mongolia.[1] These cultural realities are most acutely manifest in the conflicts that arise from an accelerating household enclosure movement. Since the national rangeland policies promote privatization and parcelization, there is both a spatial and ecological edge to the enclosure movement that needs to be explored.

## Han Spatial Identity

Early imperial China was an agricultural civilization that conceived of time and space in bounded and discrete increments, represented architecturally by the circle and square. Time was no abstract, homogenous stream, but an accumulation of definite, closed, and discontinuous periods, seasons, and epochs. Space was likewise conceived as an unending accumulation of fixed locations (De Riencourt 1958: 78). The cosmological order of time and space was maintained through court ritual that centered the universe in the emperor in Beijing, with density gradually dissipating toward the peripheries, and ultimately consumed by chaos at the frontier (ibid., 79). The self-evident superiority of Chinese culture derived from its spatial proximity to heaven, and, likewise, the geographic distance of other places to the Chinese capital determined the relative degree of civilization attributed to their inhabitants (Q. E. Wang 1999: 292).

The outer extremities of the empire were thus demarcated by an elaborate system of fortifications and walls that have captured the imagination of the world. It is extremely difficult to separate history from myth in interpreting the Great Wall and its multiple significations for the Han Chinese over centuries (Waldron 1990). Nonetheless, any reasonable starting point must concede that these massive structures have been important and dynamic ideological markers in space.

The positions of the outermost great walls have expanded and contracted through history, so that areas on the outside in one period might be on the inside in another. The system of outer walls was never a permanent or tidy barrier separating mobile herders from sedentary farmers, or even Han Chinese from northern tribesmen. The imposing barricades functioned more like a screen than an envelope, because they allowed for economic and cultural exchanges. Still, the outermost great walls of Inner Mongolia have followed, approximately, the edges of two soil zones, with the interior being arable and the exterior more vulnerable to drought, crop failure, and erosion (Lattimore 1941: 127).

The walls also clearly served as a visible ideological marker of domesticated space. In the words of Wakeman (1975: 71), "to the Chinese it marked the border between civilization and the barbarian hordes . . . that successively threatened native dynasties. To the nomads it was a barrier that challenged and beckoned." Anderson (1983: 26) once observed that premodern states typically defined themselves as cultural and political "centers" governing within territorial continuums that eventually dissolved into competing allegiances: "Borders were porous and indistinct, and sovereignties faded imperceptibly into one another." Relative to that general standard, the Great Wall system must be viewed as a rather remarkable delineation of cultural and territorial space, however permeable it may have been.

The frontier walls were not strictly military defenses but also direct instruments of agricultural extension. From the time the first Great Wall was unified during the Qin Dynasty (221–207 B.C.E.), a unique "farming-garrison" system was introduced to keep it operational. This involved the massive resettlement of civilian farmers into frontier areas, both to bolster the military garrison and to allow regional self-sufficiency in food supplies. In this way, irrigation systems were established and pasture conversion developed swiftly (Cheng 1984: 210). But the walls not only helped to extend the practices of agriculture; they also provided a forum by which to perpetuate the essential tradition of forced labor that made Chinese intensive agriculture possible in the first place (Lattimore 1941: 128–130). In this sense, the frontier walls served ideological as well as military purposes.

Nested within the frontier walls, imperial China was a land characterized by stable city walls and diminishing space. Walled cities were the major landmarks of traditional China, with a proud and distinctive morphology that, despite gradual evolution of form, remained remarkably static through history (Chang Sen-dou 1977: 100). Marwyn and Carmencita Samuel (1989: 204) have described how space was further controlled and domesticated within those outer city walls: "The Confucian city, like the Confucian house and Confucian society, was highly regimented. Its layout and structure, epitomized by a seeming endless maze of walled compounds within walled compounds within walled compounds, were imbued with the signs of power, authority, and hierarchy, and nowhere more so than in the austere formality of imperial Beijing."

Of course, the rigid cosmography of early imperial China changed in subtle ways over the centuries, but some basic themes, such as spatial hierarchy and center/periphery relationships, have proven rather persistent (Q. E. Wang 1999). The space-oriented regimentation so characteristic of imperial China has not evaporated even over the last half-century. Jeffrey Meyer (1991: 4), in describing the architectural history of Beijing, has offered the following insightful commentary on Han spatiality:

"Wall" is what makes China, wall makes the city of Beijing, the Imperial city, the Forbidden city, and all subsidiary units down to country town, village, and private home. Give any Chinese some loose bricks and he will build a wall, a gate, and hire a gatekeeper to prevent the outsider from entering.

Walls are important to the Chinese because, over and above practical consideration (preventing thievery, resisting attack, and the like), the wall is the line clearly drawn between what is significant and what is insignificant, what is powerful and what is not powerful, who is kin and who is stranger, what is sacred and not sacred. The Great Wall is the symbol of China par excellence. Traditionally it marked off civilization from barbarism; today it still marks off the "sacred land" from the rest of the world. . . .

Today walls are still a ubiquitous feature of the Chinese landscape. Even though a poor country, China lavishes an incredible amount of money on building walls where a non-Chinese would think them totally unnecessary. They are still much in favor in rural villages, and in the cities they now usually demarcate factories, businesses, schools, offices, and the other "work units" of socialist society. The Chinese passion for walls reflects their passion for clarity in human relationships, signifying an individual's identity and place within society. The Marxist revolution has in no way diminished the Chinese love of a wall. It is only that they are now built in different places, and define different units of meaning.

Even today, the cultural power of the "wall" runs deep in the national psyche. The Chinese themselves perceive the Great Wall as their greatest cultural relic and symbol (Cheng 1984: 17), though the nuances of that symbol are sometimes hotly debated. The Great Wall has alternately represented both the glory and tyranny of a Confucian heritage for centuries.[2] Under Communist leadership, it has been the object of public scorn (especially during the Great Leap Forward and the Cultural Revolution) at least as much as it has been the focus of patriotic adoration. Under restoration, it has represented the promise of a modern socialist future, and it sometimes serves as the token of an indestructible national spirit (Luo and Zhao 1986: i). There is a Chinese saying, "You cannot be considered a great man if you have not been to the Great Wall." China's national anthem enjoins the people to "build our new Great Wall with our very flesh and blood" (Cheng 1984: 7). Early in the 1980s, Deng Xiaoping hoped to inspire his countrymen along the path of economic reform with a special inscription: "Let us love our nation and restore our Great Wall" (Waldron 1990: 1).

Whether viewed in a negative or positive light, the myth of the Great Wall still maintains a powerful hold over the imagination of Han Chinese, who cherish their walls, their partitions, and their regimented space. Today, the landscape is still divided into a million interiors and exteriors. Despite the declaration of an Open Door Policy, barriers both visible and invisible dominate the landscape, perpetuating the ancient *neibu* (internal) mindset that has always separated Us from Them—Chinese from foreigners, Han from minorities, Party members from nonmembers, senior officials from rank and file.

## Han Ecological Identity

Han spatial orientations have involved an ecological counterpart. For centuries, Chinese literati viewed and described neighboring mobile peoples and their native homelands in the most disparaging terms. The people were considered to be "human-faced and animal-hearted," while the steppeland environment was "unfit for [truly human] habitation" (Waldron 1990: 38–39). Land and people were perceived in reciprocal images of savagery. "Just as their nature marked the limits of human character, their homeland was thought of as the edge of the world" (ibid., 39). The Chinese language employs numerous terms to signify the unfamiliar ecological zones of the northern frontier: *huang* (waste), *kuang* (vast), *wu* (overgrown), *ye* (untamed), *qiong* (impoverished), *xu* (emptiness). All of them are negative and convey a strong sense of malevolence. *Huang* is the most common and comprehensive single term applied. Meserve (1982: 61) has explained its subtle connotations:

Huang is "wu," land neglected, full of weeds, poor, vulgar. Huang, like "ye," is wildness and savagery, while the vast expanse that is also expressed in "kuang" adds to the menace of huang. It is land uncultivated, a land of drought and famine. Huang expresses the horror of devastation and desolation. . . . License in pleasure, total disregard of man or things, reckless excess—all are embodied in huang. It is time wasted. And huang can also mean a covering for a coffin—like "xu"—carrying with it the shadow of death.

Khan (1996: 128–129) has noted that *huang* implies moral deficiencies as well. The term suggests an absence of domestication and civility. Consequently, any land use that restructures or transforms open rangeland can only be ameliorative. "Thus, the positive term *kai* (open) is used to refer to the action of preparing a virgin land for farming: *kai huang*—to open up wasteland." This sense of pastoral depravity is further reinforced in the language by other idioms, such as *kui xin liang*. The term literally means "ill-conscienced grain." It is meant to dishonor the herdsmen who live off of grain they have not labored to produce (ibid., 140).

The power of such language has played a significant role in both motivating and rationalizing Han colonial incursions into Inner Mongolia over the last century. Migrating farmers poured into border regions under the authority of a government policy bearing the title "construct the frontier" (*jianshe bianjiang*). The migration policies gave teeth to a traditional Han perspective that any frontier lands that can be productively cultivated rightfully belong to the Han. Lattimore (1962: 417) reported that "wherever the Chinese came, the Mongols had to get out. They suddenly found themselves stigmatized as a 'backward' people, 'too primitive' to take up the new Chinese agriculture—although they had not been too primitive to take up the old 'mixed' border economy. An entirely artificial line was drawn between 'civilized' agriculture and 'primitive' pastoral economy, dependent on livestock. To be a nomad was a kind of social crime."

The derogatory Confucian attitudes were only strengthened by Marxist orthodoxy after 1949. The Marx-Lenin-Mao line of political thought held that natural rangeland has no intrinsic value as a resource because it embodies no labor.[3] Land of no value can hardly be "degraded," no matter what the manner of exploitation. To central authorities, even marginal farmland was better than natural pasture, as the "grain first" policies of the collective era continually made clear.

Beijing has tended to view the native traditions of indigenous people to be as "worthless" as the land. This political philosophy is built upon a nineteenth-century European notion of cultural evolution that held that all societies pass through a series of known stages on the way toward modernization. Hunting and gathering was the most primitive form, followed by mobile pastoralism, sedentary agriculture, and then industrial society with its class contradictions that eventually precipitate the socialist state. From this point of view, the interests of the minorities are best served by rapid assimilation (Deal 1984: 23; Connor 1984: 428–430; Tapp 1995: 198). Just as agriculture could only raise the value of the land, sedentarization could only raise the cultural level of the people.

### Mongol Spatial Identity

In stark opposition to traditional and contemporary Han perceptions, the pastoral Mongols have historically loved the open steppe and its spatial freedom. Phrases taken from a classic poem written by an ancient nomad from the northern frontier effectively capture the aesthetic sentiment of an alternate spatiality.

> As a great yurt are the heavens
> Covering the steppe in all directions
> Blue, blue is the sky
> Vast, vast is the steppe
> Here the grass bends with the breeze
> Here are the cattle and sheep
> (cited in Jagchid and Hyer 1979: 10)

The same landscape that would exhilarate a nomadic poet would drive a Han poet to despair. Honey (1992: 4) has recorded the sentiment of a Taoist sage who once left his homeland and familiar customs to seek counsel with Chinggis Khan. On his travels he became as obsessed with space as with food and clothing:

The land has no trees nor vegetation—only barren grasses. The sky produces ridges and mounds that swallow large mountains. The five grains do not mature (for food but provide fodder to) produce milk and kumiss: Now in hides and furs, and in a felt tent, I can only break out in smiles.

Traditional Mongol spatiality is rooted in a landscape characterized by mobility and mutability. Mobility is the very essence of herding. Whether on the plains of North America, the savannahs of Africa, or the steppes of Central Asia, pastoral peoples have always needed to move their animals regularly in response to the inevitable spatial and temporal patchiness of grassland resources. But herders and their animals are not the only things moving in their environment. The landscape itself shifts and moves out from under their feet as powerful forces of wind and water erosion transform the terrain day by day.

Production tied to mobility and mutability in such a direct fashion requires and instills an expansive spatial orientation. In nomadic societies, all aspects of social organization are conditioned by and subordinated to regular movement in open space. Every component of traditional Mongol culture—diet, dress, housing, labor, family form, marriage, fertility—functions in service of mobile stock-herding (Jagchid and Hyer 1979: 56; Pasternak and Salaff 1993: 170–197).

For Mongols of northern China living beyond the Great Wall, enclosed land was sometimes treated as a despised symbol of the cultivating Han civilization. The destruction of walls and other physical barriers has therefore frequently been an act of meaningful social expression. Even as imperial rulers of China, the Mongols (in sharp contrast to native dynasties and even the foreign Jin and Qing dynasties) never condoned the traditional Confucian tight regimentation of space and never themselves set about the business in any earnest fashion.

In earlier historical periods when spatial distinctions were more pronounced, Mongol assertions of power always involved the complete destruction of city walls. Chinggis Khan and his armies zealously eradicated any built structure on the landscape that was associated with settled agriculture. Grousset (1967: 245) reported how thoroughly and deliberately space was liberated under his command:

Towns were destroyed from pinnacle to cellar, as by an earthquake. Dams were similarly destroyed, irrigation channels cut and turned to swamp, seed burned, fruit trees sawn-off stumps. The screens of trees that had stood between the crops and invasion by the desert sands were down. The handiwork of thousands of years was leveled to steppe again; orchards were laid defenseless to the driving, all-penetrating sandstorms from steppe or desert. These oases . . . were nothing now but arid steppe, this by the nomads' aid making all once again its own.

The invading nomads could not imagine a useful purpose for either the agricultural populations or the tilled land that they conquered. Grousset (1967: 280–281) summarized their attitude: "Better to kill off all these useless folk who could neither tend a herd nor travel with them on their nomad migrations, better burn the harvest as they were destroying the towns, let the land lie untilled and be restored to its dignity as steppe." Mongol confiscation of agricultural land for grazing continued in North China for more than a century after conquest (Schurmann 1956: 29).

Certainly this attitude was not just driven by aesthetics. Embedded within their violent assertions of spatial preference, Mongol warriors had a clear tactical and political-economic incentive to restructure the ecologies of rival sedentary societies. Still, after successful conquest, the majority of the population preferred to remain in their traditional habitat north of the Great Wall, where they could preserve familiar social and economic relations (Khazanov 1994: 245–248). Mongol armies were more interested in extortion than colonization into unfamiliar landscapes.

Even after establishing their empire over China, the Mongol founders of the Yuan Dynasty (1280–1368) were "unsympathetic to walled city construction" (Chang 1977: 75). For a time, the Mongols prevented the Chinese from building or repairing city walls in order to display their power. Moule (1957: 13) cites Marco Polo to make this point: "When the Yuan annexed the Sung they forbad[e] the building of city walls throughout the Empire in order to display its unity, and the inner and outer walls were leveled by the inhabitants day by day." What walls the Mongols did leave standing simply deteriorated during the thirteenth and early fourteenth centuries, so that subsequent Ming rulers faced quite a task of restoration. Once the Mongols were finally expelled, the Ming dedicated the next two hundred years to rebuilding urban fortresses and the Great Wall itself, "lest remnant forces return from the north" (Cheng 1984: 7).

Even after the Yuan Dynasty, certain organizational features of Mongol nomadism helped to shape a traditional cultural system that remained firmly grounded in distinctive spatial characteristics. The Mongols practiced a Eurasian steppe variety of seminomadic (transhumant) pastoralism characterized by extensive land use, seasonal change of pasture, and supplementary production from agriculture or hunting. Their migrations were usually regular, linear, meridional, and fairly stable, with well-defined temporal schedules of movement that did not involve great distances (Khazanov 1994: 50). In regions of Inner Mongolia, the total distance of seasonal migrations rarely reached 150 kilometers (Lattimore 1951: 73). Most households migrated to the same summer campgrounds year after year, returning to an even more permanent winter location sometimes only a few miles away (Lattimore 1962: 420). Contrary to negative caricatures, they did not practice an "aimless pursuit of water and grass" (*zhu shui cao er ju*), as the Chinese popular idiom implies.

Through modern history, Inner Mongolian herders have maintained expansive spatial horizons, yet with an emerging sense of attachment to place that has always remained subordinate to suprahousehold group affiliations. During the Ming (1368–1644) and Qing (1644–1911) dynasties, Mongol herders typically organized themselves into small suprafamilial units, which consisted usually of two to twenty households that shared labor and helped to dilute environmental risks such as drought, flood, or blizzard. These groups were usually formed by agnatic kinship networks based upon mutual consensus. They were essentially residential units that formed both economic and ritual communities (Szynkiewicz 1982: 25, 32). A level above that, several of these units (usually four to twenty) coordinated their land use and access to resources informally in territorial groups sharing a common name, such as "people of one valley" or "people using the same water source" (ibid., 34). Under the Qing, Mongol nobles organized tribal peoples into banners (*qi*), which were then amalgamated into large political-administrative regions known as leagues (*meng*). This system created solid territorial identities for individual Mongols because it fixed rigid tribal borders—feudal subjects were forbidden to move out of their native banner (Bawden 1968: 109; Szynkiewicz 1982: 20).

The traditional grazing system protected the principle of open range. Though land was formally under the control of feudal lords, customary law gave common herders unlimited rights to graze their herds wherever they pleased (in coordination with regular migrations) within the boundaries of their banner, with the exception of special pastures reserved for nobility (Szynkiewicz 1982: 21; Bawden 1968: 89). In general, all of the land belonged to all of the inhabitants, so that wealth and social advancement depended primarily on the energy and competence of each individual (Lattimore 1962: 420). Though the potential for conflict over land use was great (especially during seasonal migrations), it rarely occurred, evidently because principles of proper grazing were well understood and widely practiced (Szynkiewicz 1982: 23).

After decades of social turbulence and warfare early in the twentieth century, regular organizational patterns did not resume in the grasslands until after the Communist "liberation." By 1959, most pastoral areas in northern China were fully collectivized. Within the collective structure, herding households were subdivided into smaller and smaller units of organization. The township collective (commune) was composed of brigades, brigades were composed of production teams, which in turn were composed of *duguilong* (several households grouped by proximity of residence into single production and consumption units). For two decades, the collective administration tightly regulated pasture use and almost every aspect of animal husbandry production. Collectivization resulted in fewer herding households who assumed responsibility for large, single-species herds. The remaining households worked in teams to process animal products, carry out construction, cultivate fields, or undertake a variety of utility tasks (such as eradicating wolves).

During the collective era, brigade pastures in eastern Wengniute were divided into three general grazing zones: cattle herders enjoyed the best pastures closest to residential centers, sheep herders occupied the band of second-tier pastures, and goat herders lived on the most distant scrub land. The herders remained in place after decollectivization, but the pastures formerly under their management had to be shared initially with two, three, or four other households. These groups formed the *lianhu* (cooperative households) of the reform era discussed in Chapters 5 and 7. Today, many households still express a proud sense of identity with particular pasture areas, especially those who have been rooted in one place since the collective era.

## Mongol Ecological Identity

Traditional Mongol society reciprocated the Han disdain for alternative lifestyles. Prior to collectivization, Mongol herders of eastern Inner Mongolia maintained a degree of cultural contempt for neighboring farmers, their sedentary lifestyle, and their intensive land use. For example, Jagchid and Hyer (1979: 316) have documented the Mongols' extensive style of cereal production and the local attitudes that motivated it. In Nasihan and the northern territories of what is now Chifeng City prefecture, residents used a special sickle with a long handle so that they could stand upright while cutting grass. Seeds were then broadcast by hand and the herds displaced to other pastures until fall harvest. According to Jagchid and Hyer, "this type of agriculture shows the attitude of the nomadic or pastoral Mongol who needed some agricultural products, but did not want to dig in the dirt or stoop in the back-breaking manner necessary when using the short-handled sickles of the Chinese farmer." They wanted the same cereal produce but felt compelled to preserve a separate (dignified) identity. Lattimore (1934: 77) specifically reported the hostile traditional attitudes toward intensive agriculture:

The Mongol who settled down did not do so because he felt it was a step up in civilization; he was resigned to it as a makeshift. In the same way, at the present time, the successful Mongol is the man of tents and herds. If the Mongol settles down, it is because he has been crowded by Chinese colonization until there is no room for his herds. Nothing that he gains can compensate him for the feeling of loss.

The Mongolian language, no less than Mandarin Chinese, reveals important clues about traditional attitudes toward resource utilization. Whereas the Han looked upon cultivation as "opening up wasteland," Mongol herders traditionally viewed the same activity in strongly negative terms. They called it *gajir qagalaqu*, or "shattering the land" (Khan 1996: 128). Lattimore (1934: 65) also saw in traditional Mongol vocabulary a moral perspective on the depravity of Chinese land use:

The term "hard" is used of Mongols and the term "soft" of Chinese. These terms do not stand only for physical robustness, but for the moral "hardness" of the man who lives in the saddle and makes his camp where he pleases, as against the moral "softness" of the man who is in bondage to the land he tills or the merchandise in which he deals, to his goods and his comfort, the safety of his roof and his walled town.

An especially significant point about traditional Mongolian land use patterns is the acute attention to landscape details that it required. Lattimore (1941: 242) discussed this point carefully:

"Dominant" is not really the right word to describe the influence which the lie of the land has on the life of the Mongol. "Pervasive" is better. While the site of a monastery or the placing of an *oboo* [stone memorial] may be in part a poetic expression of the way in which the life of Mongols conforms to the land, just to be Mongol means to conform to these imprecise but nonetheless valid rules. They have to do with things that are real and utilitarian, like the relation of exposure to drainage and pasture and the relation of lines of movement to areas of pasture. The rules are imprecise and relative because you yourself, the herdsman, live all your life by compromise. You have no permanently fixed point of reference in terms of "home" and "property," but you do have a succession of temporary fixed points. . . . If you are an apt herdsman, with the right "feel" for what your cattle need, it is likely to be because you have also the right feel for the lie of the land.

Of course, the Chinese are world renowned for their attention to landscape as developed in the traditional practices of *feng shui*, or the art of placement. *Feng shui* asserts that people are intimately affected by their immediate surroundings. Some landscapes are more auspicious than others, and strategic human construction can enhance or block the flow of natural forces at any particular location. As Bruun (1995: 176) has explained, "In this thinking, the environment should be utilized thoughtfully, since harmful interference hits back like a boomerang." Informed geomancy was the means to achieve stable harmony with the natural order, and every last farmer paid attention to the sacred principles.

It is less well known that the Mongols developed and practiced their own intricate system of environmental observation. In fact, Lattimore believed that Chinese geomancy itself developed out of principles first borrowed from nomadic peoples of the steppe (1941: 237). In any case, Mongol herders have historically been extremely attentive to the subtleties of landscape features, both as directional markers and as matters of practical concern. Herders must pay strict attention to the slope of hills and water, the direction of prevailing wind and prevailing sun, the probabilities of subsoil water, and other landscape features for the most obvious of reasons: if they do not, their herds will not live to see the spring. According to Lattimore (1941: 238), such attention to landscape was but one manifestation of a primal attitude towards nature: "Underlying such observances there is a genuine, sensitive and much deeper feeling that man should accommodate his needs and the use he makes of the land for himself and his herds to what one might call the needs and rights of the land itself."

Mongol herders, like Han farmers, thus traditionally nurtured their own spatial and ecological preferences. For both groups, cultural land use preferences have been rooted in the practical concerns of optimal food procurement. In arid regions, institutionalized mobility ensures the greatest access to a variety of key resources that are both ephemeral and required in differential quantities throughout the year. Though much has changed over the last century, distinctive routines, values, and perceptual thresholds still exist and inform local land use decisions.

## The Relevance of Western Assumptions

Cultural perspectives from the industrialized Western nations also play an active role in domestic environmental conflicts that arise when developing nations like China come to rely upon (or merely accept) the influences of global capital and international institutions to boost modernization efforts. The assumptions and biases of Western intellectuals can have a considerable effect upon public interpretations of environmental changes, policy goals and implementation decisions, and even the very process of scientific data collection and knowledge construction.

### SHAPING PUBLIC INTERPRETATION

Western scientific literature on land degradation almost universally regards dune sand as nothing but a menace. Sand patches are primarily characterized (in a scale-insensitive manner) as blights upon the land—pockets of deterioration that eventually "radiate out" to form expanding deserts (Nelson 1990: 17). Although desertified land is not necessarily always portrayed as mere wasteland in these scientific discussions, the notion that sand might be part of a preferred indigenous environment remains quite distant. The following account is fairly representative: "Desertification usually begins as a patch on the landscape where land abuse has become excessive. From that patch, which might be around a watering point or in a cultivated field, land degradation spreads outward if the abuse continues." (Dregne 1983: 7). Other accounts often establish a more emotional tone. For example, the 1978 United Nations Conference on Desertification asserted that "desertification breaks out, usually at times of drought stress, in areas of naturally vulnerable land subject to pressures of land use. These degraded patches, like a skin disease, link up to carry the process over extended areas" (quoted in Mainguet 1994: 1). Some accounts are even more graphic: "Desertification is not about spreading deserts. It is a rash which breaks out in patches wherever the planet's skin is mistreated" (Timberlake 1985: 60).

Curiously, book after book treats dune sand almost exclusively in the context of environmental hazard. At the more sensational extremes, dune sand is even indiscriminately demonized as a sure sign of natural (even moral) disorder (see Sears 1980; WCED 1987; Rifkin 1991). The image of a ticking time bomb best exemplifies this popular perspective: "In a little less than 200 years at the current rate of desertification there will not be a single, fully productive hectare of land on earth" (UNEP 1987: 17).

Only rarely have resource management studies in arid zones mentioned any favorable qualities associated with sand. In a significant exception, Tsoar and Zohar (1985: 184) have optimistically conceptualized active dunes around the world as the "largest potential reserves of soil" to feed an overpopulated world. They have argued that various thermal and percolation properties of sand soil create "a favourable substratum" that can support a "denser and more perennial vegetation than heavier soils such as loess" (189). Likewise, Kovda et al. (1979: 442) have reported on the potential for sandy soils optimally to retain and release moisture and nutrients to well-adapted plants.

With minimal exception, there has been almost no attention to the functional role of dune sand either among pastoral peoples in general, or among Mongol herders in particular. The flurry of research now coming out of Mongolia tends to stress the essential point that what are perceived as critical resources inevitably vary from place to place, according to season and local ecological characteristics. Besides the need for a wide variety of forage, resources that have been explicitly itemized as strategic include the distribution of water points, moist depressions, wells, salt licks, tree groves, and windward hillslopes (Szynkiewicz 1982: 20–23; Mearns 1991: 31, 1993b: 77). In one brief sentence, Mearns (1993b: 84) does indicate that sand dunes help provide relative warmth and shelter for certain communities near the Gobi. Nonetheless, to my knowledge, no research has yet reflected upon dune sand as a culturally valued environmental patch.

Yet my experiences indicate that the herders of Nasihan township exhibit a surprising degree of tolerance of, appreciation for, and even preference for dune sand at certain spatio-temporal scales. I will return to this issue in greater detail in Chapter 9. The major point to be made here is simply that influential scholars and officials within the international community tend both to ignore the fact that stock-herding populations may hold distinctive views about their home environment and to dismiss the possibility that dune sand (in certain proportions) could function as a non-threatening and valued local resource.

## SHAPING POLICY GOALS

In 1968, the American population biologist Garrett Hardin published a provocative essay entitled "The Tragedy of the Commons." Although his intention was to show that the freedom to have children leads to overpopulation (and oversaturated labor markets) and ultimately to declining environmental quality for all, the essay's great legacy was its critical commentary on common-property resource management. Hardin was concerned about societal arrangements in which benefits remained private while costs were passed on to the entire community. To make his abstract arguments more concrete, he presented the analogy of a herder grazing livestock on a public range. Each herder, he explained, would try to increase his own herd size as much as possible despite the awareness of declining pasture resources, thus demonstrating that the rational goals of individuals, if left unchecked by any higher authority, could easily lead to irrational and tragic consequences for the group. It may have been quite unintentional, but Hardin's vivid metaphor helped to popularize the view among influential intellectuals and institutions around the world that pastoral subsistence strategies were inherently destructive to the ecological environment and must be brought under control.

Hardin's essay added only incrementally to a long lineage of negative caricatures of pastoral peoples and pastoral production that inform mainstream administrative perspectives on rangeland ecology and livestock management throughout the developing world (see Scoones 1996; Fratkin 1997). Pastoralists have been denounced for maximizing their herds, refusing to sell livestock in formal markets, maintaining a diverse portfolio of animals (especially goats), resisting technological advances (such as artificial insemination to promote exotic breeds), and generally failing to "modernize" their production system. For all of these reasons, pastoral populations inhabiting dryland areas have been perceived primarily as disturbances within—rather than components of—the larger ecosystem. Yet much of this conventional animosity is based on cultural bias and misunderstanding rather than objective observation. Hostile sentiments seem rather well entrenched in agrarian-based societies. As one author has observed, "The goat embodies deep antagonisms between sedentary peasants and nomadic pastoralists that reach back to the dawn of civilization when the first Cain came to blows with the first Abel. Satan inherited Pan's horns, goatee, and hooves for reasons that remain fundamental if long forgotten. We are Cain's descendants" (Corbett 1991: 35).

Because Hardin's essay was published just before the onset (in the 1970s) of the catastrophic Sahelian Drought, in which thousands of livestock died and massive famine ensued, it was well timed to influence policy circles. His critical judgment of common-property resource management became the dominant analytical framework of international development institutions for many years (Simon 1993). The popularity of his model may be related to its peculiar ability to generate both liberal and conservative political solutions—liberals use it to assert government obligations to assume control and protect common interests, while conservatives use it to assert the virtues of privatization (Peters 1987: 171–194).

The latter (conservative) view has been more dominant, however, and many international livestock development projects throughout the 1970s and 1980s attempted to parcelize and privatize national rangelands through the medium of fence-wire. Typically, large tracts of land that once accommodated multiple uses and various users became partitioned and distributed among single or small cluster households, always with highly significant consequences for land use and local social relations. New restrictions of access to land resources have been widely conceded to introduce troubling social inequities, but policy makers usually justify this result as the necessary cost of good stewardship. As Hardin (1968: 1248) so succinctly put it: "Injustice is preferable to total ruin."

Yet the routine disposal of common lands and common herders does not always enhance prospects for sustainable land use. As recent government initiatives come of age, evidence mounts that privatization is no tidy solution to land management problems. Anthropologists, in particular, have labored at length to demonstrate the many oversimplifications inherent in Hardin's interpretive framework (see McCay and Acheson 1987; Bromley and Cernea 1989; Feeny et al. 1990). In the last decade, numerous studies have stressed the general inapplicability of the model to traditional societies, demonstrating from ethnographic research that indigenous forms of land utilization do not in fact inevitably lead to range degradation (see Goldstein et al. 1990; McCabe 1990; Buzdar 1992). Other field studies argue conversely that privatization does not exactly guarantee conservative rangeland management (Sandford 1983; Little and Brokensha 1987). A third point of critique has been to show that the legal tenure system is itself less significant than other economic factors. Once a supportive production climate is created, agricultural yields and resource management both improve, no matter whether the land is held as a private or collective asset (Simon 1993; Cousins, Weiner, and Amin 1992).

Thus, there is by now considerable ethnographic evidence to support the position that common property resources are not inherently problematic. In any given location, common property status does not necessarily lead to, nor suffice to explain, the event of resource depletion. Yet policies that are based on the flawed assumption that common property arrangements belong to an outdated earlier stage of social evolution continue to be adopted all over the world.

## SHAPING SCIENTIFIC KNOWLEDGE

One of the most pervasive and ingrained philosophical axioms in Western intellectual history is the principle of dualism, which assumes that reality can be analytically divided into two broad categories, namely the apparent distinction between matter and mind (or body and soul, form and substance, nature and culture). This foundational premise accounts for the very organizational structure of Western science—those rigid walls of separation that still divide the sciences from the humanities and the natural sciences from the social sciences—and imposes cultural and institutional biases of great subtlety and consequence for the construction of rangeland science and resource management policies.

Over the last several decades, there has been a marked increase in institutional emphasis upon the need for interdisciplinary collaboration to address pressing environmental concerns, including deforestation and land degradation in dryland areas. Despite the explicit attempt to break down conceptual walls of disciplinary separation, it remains doubtful whether such projects have actually escaped the old conventions and intellectual dichotomies (see Hjort 1982; Shipton 1994). The immense frontier beyond that metaphorical "Great Wall" remains largely unexplored.

A fundamental obstacle to interdisciplinary collaboration is the absence of a common framework to talk about human–environment relations in nonconfrontational terms. The language of scientific analysis still requires privileging either nature or culture as a dominant force of environmental transformation. Even research that shares a common focus upon landscape ecology cannot get beyond the conceptual straightjacket (Bohm and Peat 1987; Naveh and Lieberman 1990). The intellectual division of labor between physical and social scientists has been especially rigid in the practice of Chinese grassland ecology, where social factors in ecosystem dynamics have been almost totally ignored (Loucks and Wu 1992: 80).

These realities have direct implications for the production of scientific knowledge in grassland communities. In the case of the GEMS project, disciplinary identities of participating scientists helped to channel them into rather conservative research alliance strategies. For the most part,

Western scholars collaborated with Chinese or Mongolian scholars from the same or closely related disciplines. Rarely did collaboration occur between natural and social scientists, as originally intended. And when it did occur (as in my case), collaboration tended to be more logistical than cognitive.[4] Thus, the structure of Western science helped to structure the kind of collaboration that could occur, the kind of research questions that could be asked, the kind of data that could be collected, the levels of funding that could be expected, and the channels through which scientific information eventually could be disseminated. The damage to scientific discovery follows primarily from the fact that the ideological and institutional tensions that exist between natural and social scientists within an international frame of reference help to conceal the ideological and institutional tensions that exist between urban Han and pastoral Mongol within a national and local frame of reference (see Williams 2000).

The invisibility of such cultural bias makes it easy for the Western scientist to be unaware that alternative representations of nature even exist in Inner Mongolia. The structure of engagement with local data generally compels them to endorse rather than challenge the Chinese discourse concerning the causes and culprits of land degradation and the policies considered necessary to control them. I witnessed this process in operation, as the research station hosted many international delegations during my year of residency.[5] I was astonished that delegation upon delegation verbally endorsed the Han perspective and the full range of national grassland policies by the time they left the research station. Their comments generally conformed to the sentiment explicitly communicated by one European scientist that "the work of the research station perfectly matches the needs of the local community."

Natural scientists from Western nations and Japan almost invariably rely upon Han scientists for access to practically all field data. They do not speak local languages and so receive their critical orientation primarily through the filter of translation. They do not stay long enough to explore beyond a predesignated tour route. They gain no access to those in the community who dispute the research station's version of reality. Even if they do come across such people, they hear nothing unorthodox for want of time to establish a trusting relationship. But they usually do not seek out local inhabitants, because they perceive no need. Western scientists naively consider their hosts to be the "local experts," even though Han scientists see themselves as outsiders who work in an alien environment among alien people. Undoubtedly, Han scientists do develop an identifiable expertise through their laborious field research. I am not arguing that all of their data is contaminated or that they have no valid perspective by which to interpret ecological change. But their experiences and knowledge base must not be construed as "local," "insider," or "native," given Inner Mongolia's colonial history.

It will be possible to return to this discussion in the concluding chapter, but for now, the essential point is that the sustained intellectual dichotomy between nature and culture carries adverse practical consequences for scientific investigation. At a minimum, it allows distortions of data to go undetected, it conceals the linkages between science and social power, and it prevents dialogue between competing knowledge systems that must occur in order to improve understanding of environmental transformation.

# FIVE

# The Community: History, Production, and Residual Fear

An ethnographic exploration of social realities at the community level provides us with essential details that act as a critical counterweight to scientific understandings based primarily upon more distant and detached surveillance strategies such as remote sensing, rapid rural appraisal, and other forms of grassland tourism. The township is the general unit of analysis, but it also will be useful at times to provide some details in the context of a single village. Thus, the discussion will occasionally shift focus between Nasihan township and Wulanaodu village.

## *Historical Background*

### PREHISTORY THROUGH THE REPUBLICAN ERA

Chifeng City prefecture was the ancient location of a famous Neolithic settlement known to archaeologists as Hongshan Culture. This culture practiced settled cultivation in the region from 4000 to 3000 B.C.E. Later Neolithic cultures also practiced cultivation in the area until a sharp decline in temperature and moisture patterns began around 1000 B.C.E. From that time, cultivators and stock-herders began to occupy separate territories along a fluctuating precipitation gradient that defines the Mongolian steppelands.[1] The sand lands that run along both banks of the Xilamulun River today are the residual product of these earlier climatic trends (Reardon-Anderson 1995: 29, 35).

Much later, the land around the Xilamulun River formed the heartland of the nomadic Kitan peoples who founded the tenth-century Liao Dynasty (Jagchid and Hyer 1979: 11). In their southern military conquests, they relocated hundreds of thousands of farmers to the grasslands. With the security of surplus grain provided by the farmers, pastoralists increased the size of their herds while shifting their composition from sheep and cattle to include horses and camels. The force of grazing more and larger stock on the rangelands, combined with a sustained climate shift toward colder temperatures and less precipitation, brought changes to the environment that contributed to dynastic ruin (Reardon-Anderson 1995: 39).

Beginning in the twelfth-century Yuan Dynasty and continuing into the fourteenth-century Ming, the return of a more favorable climate once again permitted the expansion of agricultural practices deep into the Chifeng region of Inner Mongolia. By the mid-eighteenth century, Chinese cultivators, then under Manchu control, penetrated northward into Wengniute banner and even beyond the Xilamulun River, successfully raising wheat, barley, buckwheat, millet, and even rice for a time (ibid.: 55). The migratory flow really accelerated when the young Chinese Republic declared in 1914–15 that all Mongol lands belonged to China, and that Mongol land titles were invalid unless ratified by provincial-level Chinese authorities (Jones 1949: 61; Lattimore 1934: 103–106).[2]

In Wengniute, Han incursion was synonymous with agricultural expansion. Migrants forced mobile stock-herders onto increasingly smaller and more marginal tracts of land reserved for them toward the east. Even today, the banner is spatially and socially divided into a western agricultural zone (with better soils and more efficient watershed), and an eastern pastoral zone characterized by moving dunes. The west accounts for roughly 53 percent of banner land mass, 91 percent of cultivated fields, and 90 percent of the population (Reardon-Anderson 1995: 64–65). With the steady flow of migrants, local officials lost the ability to conserve dwindling forest cover from commercial exploitation. Tree felling operations only intensified after the Japanese occupied the region beginning in 1933. Banner officials, however, report that most of this deforestation occurred in the west, where tree cover had been greatest.

The era of Japanese occupation is particularly instructive for what it reveals about Mongol responses to policies of technological intervention. The Japanese were anxious to improve the native Mongolian breeds of sheep and quality of wool in order to replace the Australian source of supply for their domestic woolen industry. They introduced new sheep breeding and rearing methods, as well as veterinary services, but these benefits were negated by the Japanese wool-purchasing monopoly, which expected Mongol herders to sell at fixed prices that were too low to gain a reasonable profit (Jones 1949: 66). In Japanese-occupied areas, the livestock "improvement" program was intent on breaking down the Mongols' subsistence economy (Lattimore 1962: 431–432). The refined quality of wool would mean a decrease in the hardiness and meat value of sheep, as well as a decline in the local utility of wool for making durable felt. The introduction of refined breeds would also result in Mongol dependence upon Japanese exports for food, clothing, and housing. Lattimore (1962: 432) observed:

It is the conservative stubbornness of the Mongols in wriggling away from the control of a money economy which, more than difficulties of climate and pasture, accounts for the Japanese failure to increase rapidly the herds of "improved" sheep in Manchuria. . . . The "ignorant and backward" Mongol prefers a relatively low economy, under which he is his own master, to a relatively high economy under which he would become the coolie employee of Japanese wool-growers, dairy interests, and cavalry-remount breeders.

The conservative response of the Mongol herders to Japanese technological solutions seems rational enough given Lattimore's portrayal of a Faustian bargain. It provides some evidence of the persistence of traditional Mongol values of independence, mobility, and self-determination.

The long years of World War II brought a varied stream of occupying armies into the region. Japanese, Soviets, Chinese Nationalists, and the famous Eighth Route Army of the Communist "liberation" forces all wandered through Wengniute, feeding off the animal produce. Indeed, there is still a pile of rocks in one Wulanaodu pasture commemorating the site of a battle in which some twenty Communist soldiers died in ambush at the hands of hostile Mongol herders who did not appreciate the confiscation of their livestock. It was just such constant pilfering that led an infamous Nasihan resident—Han Cangjie—to create a Bandit Federation that grew to about 1,000 men and controlled territory over three counties until 1947. A useful combination of sand, marsh, and thick willow tree cover made the area an ideal warlord hideout. Over the years of turmoil from roughly 1930 to 1950, the animal population declined significantly in Wengniute and eastern Inner Mongolia, only returning to earlier stocking levels in the mid-1960s.

## MAOIST ERA

The Eighth Route Army initiated land reforms in the area even before the end of the war. Starting in 1947, livestock, land, and even household possessions were taken away from landlords and more prosperous families to be distributed among the poor. The process was horribly violent. Many residents of Nasihan testified to severe physical beatings that sometimes led to death or permanent injury. One informant described the chaos of that period by recalling an experience he witnessed while traveling through the countryside:

In 1947, the Communist party policy was fervently dedicated to "social rectification" (*zheng feng*). Most people were caught up in the appeal for class struggle. Mobs used to take their revenge against rural landlords by tying them to the tail of a horse and then sending the horse into fields where the sharp stubble of harvested sorghum (*gao liang*) stalks would beat and cut them to death. They would also take a hot brand and mutilate the genitals of female landlords. These were common forms of punishment among the Han, whereas Mongols tended to suspend people in the air and beat them to death.

Another informant testified that Han outsiders from neighboring agricultural communities came to stir up local hatred against long-standing residents. "Herders, in their fear, would lie about the size of their own herds and credit them to a neighbor, who would then receive a beating and lose all possessions." In this way, relatively prosperous families were replaced by new brokers of social power.

Nor did the national campaigns of terror subside after the initial reforms. In 1951, a campaign swept the countryside to suppress all those who had opposed the Communist takeover. Many residents of Nasihan who survived land reform were beaten to death at this time. In 1952, a campaign to struggle the masses against "three vices" (corruption, waste, and insubordination) perpetuated the violence. In 1953, a campaign to struggle "five vices" targeted new and old enemies of class, ideology, and behavior. In 1956–1957, an "anti-rightist" campaign visited gratuitous violence upon millions of intellectuals and community leaders. In 1958, a "mop up" campaign to search out remnant "enemies of the people" again mobilized the rural masses in a political witch hunt.

In 1966, the Great Proletarian Cultural Revolution unleashed a frenzy of violence on a scale and magnitude surpassing that of the initial land reforms. In IMAR, a campaign of paranoia (labeled the "anti–New Neirendang" campaign) directed specifically against ethnic Mongols resulted in great persecution and slaughter. Rumors began to circulate widely that the Neirendang (Inner Mongolian People's Party) had never dissolved during the Communist "liberation" as most people thought, but had secretly been converted into an underground Mongol party that infiltrated local government circles in order to promote the secession of Inner Mongolia from China and its reunification with "Outer" Mongolia. It was presumed to be supported by the Soviet Union, and all good citizens of

China were called upon to help root out its members. The estimated death toll ranges from 10,000 to 100,000, while some 300,000 Mongols were accused of fomenting independent nationalistic aspirations (Jankowiak 1988; Sneath 1994: 422). No one was above suspicion, and those accused were expected not only to confess but also to name their accomplices, thus feeding the paranoia and frenzy. In reality, the party that Maoist zealots imagined did not even exist at the time (Jankowiak 1988; Sneath 1994: 420).

There is hardly a resident of Nasihan who does not have some tale of anguish to relate from these troubling years, either about themselves or a close relative. Several residents took their own lives. I consider it instructive to report briefly the events of one personal account as it was told to me by surviving family members and friends.

In 1956, a young man named Hasibagan worked as an assistant principal in a middle school near the banner seat of Wudan. One day, the local newspaper ran a story about Ulanfu, then governor of IMAR and revered by most ethnic Mongols, but printed it along with his picture on page two. Hasibagan observed aloud that this action was inappropriate, pointing out that a newspaper would never run a story about Chairman Mao on page two. Hasibagan felt that, given Ulanfu's status as the leader of Inner Mongolia, he no less than Mao deserved first-page treatment. The principal of the school was an ethnic Han who took issue with the remark and later reported the discussion to higher authorities, asserting that Hasibagan had idealized Ulanfu as equal in significance to Chairman Mao. With this accusation, Hasibagan was labeled a "rightist," fired from his position, and banished (*xia fang*) to the township of Nasihan.

Hasibagan spent the next twenty years hoping for political rehabilitation. When Mao finally died in 1976, however, Hasibagan gave up hope of achieving any closure or restitution for past injustices. He devised his own political rehabilitation through symbolic suicide. The newspapers announced the exact day and time at which a national memorial tribute to Mao would be organized. Sirens and horns would sound all over China for a full three minutes, after which every citizen was to bow three times toward the image of Mao at Tiananmen Square in Beijing. With this information, Hasibagan calculated the time it would take him to die after drinking a cup of insecticide. He then synchronized the lethal act to coincide with the moment of tribute. With that gesture, he apparently considered himself both participant in and object of national mourning.

The 1950s brought rural collectivization. This involved a momentous change in both social organization and local ecology. In this region of pastoral China, widely dispersed herding households first had to be settled and organized into residential districts. Until the early 1950s, the population remained sparse and transhumant. Lattimore (1934: 263), for example, described the region as "not so thickly settled by Chinese because the land is poorer. . . . From Wudan eastward, in the angle between the Xilamulun and its main southern affluent, the Laoha, there is actually a good deal of open pasture remaining, in which Ongniod [Wengniute], Aohan, and Naiman Mongols keep up the nomadic life."

Contemporary Wengniute officials in the central city of Wudan claim that grass and tree cover in the pastoral eastern zone of the banner remained relatively stable until the two great campaigns of the Maoist era— the Great Leap Forward and the Cultural Revolution—took their toll on the environment. A number of local informants corroborated the opinion that moving dunes occupied only a limited area within Nasihan itself prior to the 1950s. (One especially credible source who lived in the Wulanaodu region as an adult for almost fifty years estimated that in 1950, sand covered only 20–30 percent of the land surface.) Such discussions, however, left the distinct impression that these assertions were not intended to convey accurate geographic information so much as to perpetuate a local narrative that resources were pilfered and consumed by outsiders, primarily during the periods of Maoist excess. Historical sources, by contrast, indicate that desert sands dominated much of the landscape in the triangle between the Xilamulun and Laoha rivers since at least the turn of the twentieth century. Hedley (1906), during his missionary travels to the region, passed just to the east of what is now Nasihan township, along the Laoha River, and described the environment in this way:

The district is nicely wooded, with poplars and willows only, but the land is entirely sandy. Hence when the wind blows, the sand flies . . . Wind! Wind!! Wind!!! Dust! Dust!! Dust!!! . . . The way before us to the junction of the Xilamulun and the Laoha leads over heavy sand-hills, where no tracks can be traced, and where, unless we have a guide, we are likely to find ourselves in great difficulties. . . . The land along which we have traveled is largely uncultivated and the population is very sparse. There is now but little timber about, and so we are in the regions where cow manure is used for fuel. . . . Look north, south, east or west, as far as the eye can reach, one can see nothing but sand-hills of varying form and height, stretching away on both sides of the river. . . . The slightest breath of wind lifts the light and delicate sand, and in a few hours will obliterate the most distinct path. The wind is also constantly causing the hills to change their form, so that there are practically no permanent landmarks by which the way may be traced. (81–86)

Another eyewitness account comes from Broughton (1947: 22–23), also a missionary who spent time in a Japanese prison in Chifeng but passed through Wengniute. Describing an elevated pass above the Xilamulun River, he wrote, "I saw a view which was awe-inspiring in its majestic yet barren splendor. In the foreground were giant sand dunes, such as I had never imagined existed. They extended to the distant east, like waves of the sea, becoming smaller as they spread out towards the Manchurian plains. To the north and west were range after range of rugged hills and grassy moorlands." Both these accounts are less than precise about the geographical location of their panoramic view. Still, they indicate that a high ratio of moving dunes to stable grassland in the vicinity of Nasihan has defined the terrain for quite some time. Even as early as the Tang Dynasty (618–907), the Xilamulun River basin was designated by officials of the Chinese court as the "land of pine trees and desert" (Jagchid and Hyer 1979: 398). And despite its obviously propagandistic tone, the account of two journalists who visited Nasihan in the early 1960s indicates that mobile dunes have probably pervaded the area for at least a hundred years (Manduhu and Nasendelger 1963).

Clearly the moving dunes that dominate the landscape of eastern Wengniute today have not all emerged just within the last four decades. Nonetheless, that fair assumption does not contradict the fact that intensified anthropogenic pressures dating from the mid-1950s have contributed substantially to dramatic ecological transformations that long-term residents find disturbing. Specifically, the Great Leap Forward era diverted an enormous amount of local resources to the construction of a huge reservoir in Hongshan whose value to the region was at least as much symbolic as it was economic. The intended economic benefits were flood control and electricity generation. The unstable river basin had long been a hazard for agricultural production. Hedley (1906: 44) reports that before the reservoir, residents feared the Laoha River, describing it as "just as uncertain and treacherous as the Yellow River itself, China's sorrow par excellence." One especially disastrous flood occurred in 1883, which destroyed crops on both banks for many miles, and drowned many residents. Mao arranged for Soviet experts to help construct the reservoir, though relations chilled and the Soviets abandoned the project halfway through. Over and above the intended economic benefits, the reservoir was a colossal symbolic statement addressing many concerns of the day: the effectiveness of China's new government, benevolent intentions in national minority areas, and Sino-Soviet cooperation. Perhaps the greatest symbolic value was the conspicuous display of a government structure that tamed the barren wilderness and imposed upon it the seal of Han civility—rice cultivation.

The Hongshan Reservoir was constructed from 1957 to 1962 with an estimated 100,000 workers who disposed of some 30,000 cartloads of hardwood trees (Reardon-Anderson 1995: 77). Boulders were tied to tree trunks and submerged to build a temporary coffer dam. In addition, the workers consumed several tons of firewood each night for warmth and cooking fuel. Local herders, eager to earn a small sum of cash for every fifty kilograms of firewood they could haul, ransacked their own environment on behalf of the construction work force. Wooded areas within a fifty-kilometer radius of the construction site were depleted (ibid.: 78).

I have already discussed the disastrous ecological consequences of the Cultural Revolution in terms of its "grain first" policy, the loss of productivity, and the atmosphere of fear it created. With regard to specific impact on local resource management, the Cultural Revolution unleashed an era of unrestrained consumption of local wood for fuel, again depleting timber stands that had made something of a comeback after the construction of the Hongshan Reservoir. Old-timers from the community today estimate that even after reforestation efforts were initiated by the grassland research station in 1975, less than 10 percent of previous forest cover has been restored.

Another characteristic problem was that local resources and labor were wasted on many collective projects simply because everyone was afraid to question orders from above. In Wulanaodu, a production brigade invested twenty laborers every day for six months in an effort to dig a huge well that was, in fact, too large to be practical. The men knew as they worked that it could never be used, but no one dared bring it to the attention of higher authorities. Today it remains an unfinished and unguarded hazard for wandering animals who occasionally fall in and drown.

In Mongol areas, most of the violence of the Cultural Revolution was conducted on the authority of a single phrase from Chairman Mao: "destroy the old ways and construct the new" (*puo jiu li xin*). This slogan brought destruction to temples and religious shrines throughout IMAR. In Nasihan, residents report that Red Guards demolished every *oboo* (stone memorial) on every hilltop to suppress the traditional rituals and spiritual aspirations of devout patrons. More generally, the Cultural Revolution and its memory repressed the free expression of ethnic identity in the region for nearly two decades. It also taught thousands of pastoral households to survive political turbulence by practicing routine survival strategies of feigned compliance and passive resistance. This bitter mind-set still has consequences for daily decisions about how to manage land and other resources. It helps to explain many otherwise curious herder attitudes and behaviors that exhibit great caution and apprehension in the face of economic risk and opportunity since decollectivization. One resident elder often explained puzzling local conduct to me by slowly and dramatically drawing in the sand four Chinese characters that translate as "the heart has residual fear" (*xin you yuji*).

## Contemporary Circumstances

### DISTINGUISHING CHARACTERISTICS

Wulanaodu village made a mark in Chinese history as one of the first pastoral communities to fully collectivize. The community was officially founded in 1954 when a Party official named Gengden was transferred from neighboring Haersu city with a mandate to round up dispersed homesteads and organize a herding collective. At that time, the village was named Zhaoketu and consisted originally of only twelve households. By 1958, with some forty households, Gengden had successfully collectivized all assets and won national recognition as a model pastoral commune. Mao himself praised the village in a speech from Beijing and ceremoniously awarded it a symbolic red star of merit. The honor was later commemorated in an award of recognition dated December 1958, which bears the stamped seal of Zhou Enlai. Afterward, the community decided to change its name to Wulanaodu, which in Mongolian means "red star."

Public recognition of collectivization made the village regionally famous and helped entice other pastoral households to settle in the area. The fame later served to gain the community certain advantages, such as subsidized fence-wire for the purpose of setting aside reserve meadows for winter hay production. Roughly 400 hectares of the most fertile pasture land was enclosed by the collective between 1964 and 1977. It was in large part a result of this subsidized wire that some households had the early opportunity upon decollectivization to seize it for their own private use.

Wulanaodu fame influenced village access to other resources as well. In 1959, a delegation from the People's Republic of Mongolia toured North China and made a stop in Wulanaodu. Regional authorities were embarrassed to reveal that the renowned commune still had no electricity, so state funds were provided to jury-rig a system that would temporarily yield electrical power until the foreign guests departed. After one month of use, the commune dismantled and returned the equipment, waiting another twenty years until electricity finally became available (at least within the village residential center).[3] In homes built on outlying rangeland (accounting for slightly more than 50 percent of village households), electricity is still unavailable today.

## CLIMATE

Nasihan is located at a cold northern latitude of 44 degrees. The agricultural season is short, with only 130 to 150 frost-free days per year, and a yearly average of only about eight hours and twenty minutes of sunshine per day (Xu Lan et al. 1990: 235). The yearly average air temperature usually registers between four and six degrees Celsius, with daily winter temperatures often falling lower than thirty degrees below zero. From 1957 through 1990, each month of the cold season delivered the following subzero mean temperatures (Celsius):

| | |
|---|---|
| November | −2.7 |
| December | −9.8 |
| January | −12.4 |
| February | −9.9 |
| March | −2.1 |

Cold regional temperatures are compounded by extreme winds, which obtain yearly mean speeds of 4.0 m/sec in Nasihan (sand becomes airborne at a wind velocity of 5.0 m/sec). Furthermore, wind speeds achieve high gale force levels up to seventy days per year (ibid.). There is a local expression about the wind that jokes that there is but a single wind every year that blows from spring through winter (*mei nian you yi chang feng, cong chun gua dao dong*). During winter months, the implications for heat loss from wind chill can be quite serious.

The climate is also arid, with mean annual precipitation usually ranging 300–500 mm. The 1957–90 yearly average amounts to only 368.8 mm, whereas yearly mean evaporation reaches 2,200 mm. Table 5.1 indicates average precipitation for each year from 1957 through 1990, as measured in the banner seat of Wudan city. Table 5.2 indicates total cumulative and average precipitation in the same area for each month from 1957 through 1990.

In summary, inhospitable natural conditions of sunlight, temperature,

TABLE 5.1

*Yearly Precipitation, Wudan City, 1957–1990*

(mm)

| Year | Rainfall | Year | Rainfall | Year | Rainfall |
|------|----------|------|----------|------|----------|
| 1957 | 412.8 | 1969 | 486.6 | 1981 | 241.7 |
| 1958 | 342.8 | 1970 | 328.1 | 1982 | 347.7 |
| 1959 | 563.9 | 1971 | 269.4 | 1983 | 355.6 |
| 1960 | 313.3 | 1972 | 293.6 | 1984 | 403.3 |
| 1961 | 272.8 | 1973 | 300.1 | 1985 | 455.7 |
| 1962 | 331.7 | 1974 | 503.3 | 1986 | 519.3 |
| 1963 | 387.0 | 1975 | 397.6 | 1987 | 425.1 |
| 1964 | 508.7 | 1976 | 371.8 | 1988 | 228.6 |
| 1965 | 280.1 | 1977 | 387.1 | 1989 | 261.9 |
| 1966 | 479.2 | 1978 | 372.9 | 1990 | 452.2 |
| 1967 | 250.6 | 1979 | 416.0 | AVG. | 368.8 |
| 1968 | 287.5 | 1980 | 289.9 |  |  |

SOURCE: Unpublished weather station statistics, Wudan, Wengniute.

TABLE 5.2

*Monthly Cumulative Total and Average Precipitation,*
*Wudan City, 1957–1990*

(mm)

| Month | Total | Average |
|-------|-------|---------|
| Jan. | 47.9 | 1.4 |
| Feb. | 88.6 | 2.6 |
| Mar. | 196.5 | 5.8 |
| Apr. | 365.5 | 10.8 |
| May | 966.7 | 28.4 |
| June | 1,985.5 | 58.4 |
| July | 3,814.9 | 112.2 |
| Aug. | 3,066.4 | 90.2 |
| Sept. | 1,352.1 | 39.8 |
| Oct. | 485.3 | 14.3 |
| Nov. | 133.5 | 3.9 |
| Dec. | 35.6 | 1.0 |
| FULL YEAR | 9,538.5 | 368.8 |

SOURCE: Unpublished weather station statistics, Wudan, Wengniute.

wind chill, and aridity combine to create a harsh and dangerous work environment. As a group, the Mongols have historically suffered terribly from rheumatism and other climate-related illnesses. They are especially susceptible to health risks of chronic and acute cold stress, as well as accidental injury and death from hypothermia for at least six months of every year.

FIG. 5.1. Thematic Mapper (TM series 5) remote sensing photo of Nasihan township, September 1993, with Wulanaodu village positioned at center

## PHYSICAL ENVIRONMENT

Figure 5.1 shows remote sensing data for Nasihan township from September 1993. Wulanaodu village is positioned in the middle of the photo. In a color print, sand appears white, light grass cover appears green or brown, the reserve meadows and cultivated fields appear red, and water appears black. This image will provide a point of reference for discussion throughout the book.

Nasihan has a total area of 613 square kilometers, whereas Wulanaodu village has a total area of 88.61 square kilometers. According to 1992 population statistics, Nasihan population density is only about 6.45 persons per square kilometer, whereas the figure in Wulanaodu is more like 8.2. Local vegetation is a transitional variety from forest to steppe. A remnant forest of pine and oak still exists some 20 km from the research station. The production landscape is rather varied. For example, the total land area in Wulanaodu breaks down roughly as follows (all figures are given in km$^2$):

| unusable land | 14.7 |
| grazing land | 66.7 |
| woods | 6.8 |
| arable land | 0.4 |
| pond area | 0.13 |

Nasihan landscape, despite its variety, is dominated by moving and semi-fixed sand dunes that wind and water erosion has shaped into a jagged terrain.

According to the publications of the research scientists at the Grassland Ecosystem Research Station in Wulanaodu, the region has four major types of sand soil (Wang et al. 1984; Xu Lan et al. 1990). The first is mobile-dune sand soil, which accounts for roughly 15 percent of the total land area. At the crown, dune height fluctuates up to a meter per year, although under extreme conditions it might vary by as much as 80 cm on a single day. The dunes advance at a rate of two to three meters every year, although more recent observations indicate an accelerated rate of five to ten meters per year (Nan et al. 1993). The second soil type consists of semifixed and fixed dunes, which together account for another 15 percent of the total land area. Semifixed dunes have vegetative cover in the range of 30–70 percent, while fixed dunes are characterized by at least 70 percent cover with subsurface moisture. The third soil type is grassland sand soil, accounting for about 27 percent of the total land area. The fourth is alkalized meadow sand soil, accounting for about 14 percent of the total land area. (The remaining land area consists of woodlands, ponds, pasture, and arable fields.)

The terrain makes mobility extremely difficult while paradoxically ensuring the need for daily long-distance travel, since impoverished land resources necessitate extensive land use. Except within the heart of village residential centers, neighbors tend to be few and far between. All local traffic between private homes and village centers moves on horseback, mule litter, or human legs, across shifting dirt paths and sand traps. Only a few government officials have access to a bicycle, and a single motorcycle and jeep are shared among the highest level township cadres. A few entrepreneurs are pooling money to purchase motor vehicles, but these are still rare. Travel routinely takes up a good bit of time and energy every day, given the topographical difficulties of movement, in combination with a production regime that demands mobility for even the most ordinary routines of labor, trade, and sociability.

## MARGINALITY

The familiar core–periphery model often serves as a useful framework for understanding exploitative social realities within national border regions like Inner Mongolia. The monolithic concept of peripherality can be usefully broken down into four distinct dimensions: geographic, cultural, economic, and political (Rokkan and Urwin 1983: 2–6). From the perspective of Beijing, Nasihan constitutes a marginal community in every dimension, and each dimension conditions levels of income, access to capital, nutrition, hygiene, and government services of health and education.

Geographically, a 400-kilometer separation from Beijing might not seem so distant until one considers the turbulent history that has both connected and divided pastoral Mongols and agrarian Han peoples for centuries (see Barfield 1989). Nor have the steppelands ceased to function in their historical role as host to geopolitical border conflict. Wengniute is strategically poised to be a battlefield in the event of a land invasion by Russian armies in the north. This threat has subsided in recent years, but there are plenty of military landmarks in the region that underscore both the reality of Cold War tensions and the sense of geographic marginality relative to Beijing. Throughout regions of Wengniute, tanks lie buried in sand traps, pill boxes adorn many hillsides, and bombing targets for pilot training dot the countryside. In Wulanaodu itself, a military "iron triangle" towers atop the highest hill in the area, providing a reference point for mobilized soldiers. Reserved hay meadows exhibit "irrigation channels" that do not successfully convey water, and were never meant to because they were constructed to be an underground garrison. At the height of the Cold War (in the mid 1960s), commune leaders received instruction to prepare a place to hide weapons and ammunition in the event of Soviet aggression. These landmarks were never utilized, but they further reinforce the message of risk and fear established by so many built structures upon the landscape.

Cultural marginality manifests itself in numerous ways. In urban areas of IMAR, large-scale migrations over the last hundred years have confused the tenuous footing of social identity for both Mongols and Han, who generally don't know where to find themselves among the public display of scrambled ethnic symbols (see Jankowiak 1993: 8–96). In rural areas, however, the contrasts of language, lifestyle, and sentiment remain more palpable and tend to reproduce segregational impulses. In Nasihan, the total population is 95 percent ethnic Mongol, within a banner, municipality, and province composed of Mongols at rates of 14 percent, 12 percent, and 14 percent, respectively. Cushioned within an ethnic pocket, many Nasihan residents lack the confidence, social skills, and resources to interact advantageously with surrounding Han-dominated areas. Stories of mistreatment related by returning migrant laborers in recent years only feed widespread anxieties about the encroachment of the threatening world beyond township borders.

Economically and politically, Nasihan remains isolated and remote, a peripheral zone of the world economy where production remains low and surplus is largely siphoned away by the price scissors of high-input costs and low-output returns. The mean net income for village residents in 1992 was only 400 yuan (U.S.$50), compared to 695 yuan for all banner herders, and 1,874 yuan for banner industrial workers (Wengniuteqi renmin zhengfu 1993: 5). Wengniute is officially designated a "poverty county," and thus entitled to central government assistance, as well as provincial-level aid. What is given with one hand, however, is taken away with another as government investment in local support services for animal husbandry development remains low. The "extension services" available in Nasihan are limited to overpriced livestock nutritional supplements and limited veterinarian supplies, which can be purchased if any personnel happen to show up for work. I made the trip in vain many times.

The slow pace of bureaucratic services further testifies to the political marginality of Nasihan residents. For example, a Wengniute government committee issued a new policy statement regarding fines to be assessed for family planning violations on January 1, 1993. But the news did not reach village residents until mid-May. Also related to this problem, residents complained to me frequently about long delays (stretching to years) in obtaining a variety of state financial obligations due them. A rudimentary transportation infrastructure perhaps bears the greatest testament to the region's ongoing marginality. A single unpaved road made of rough quarried stone cuts through the middle of Nasihan, tenuously connecting residents to distant market centers only since 1966. Bus service did not become available until 1973, and sudden showers or extreme wind and temperature still frequently impede traffic flow.

Each of the four aspects of Nasihan's marginality directly conditions regional living standards. Low levels of disposable income translate directly into poor-quality housing for most residents. Most common are earthen homes, which develop cracks that require mending every spring, until the whole house collapses after about ten years. With no insulation and no central heating, chronic cold stress may develop even indoors by virtue of draughts and unstable temperatures, especially during hours of sleep when hearth fires expire. Low income also limits access to optimal nutrition, education, and health care.

## SETTLEMENT PATTERN AND LABOR GROUPS

Until decollectivization, local authorities carefully restricted where the population could settle and build homes. Only families charged with the daily responsibilities of herding collective livestock could settle out on the range. All other households were expected to live in the designated residential district located near the highway. Over several decades of spatial segregation, some distinctive differences have developed between the two settlement communities. The most important is that pasture residence involves much less social contact, which influences many aspects of social life, including educational opportunities, work habits, and mental attitudes. In the pastures, there is no electricity, few residents speak Chinese well, diversions are few, and parameters of time and space are expansive. Despite many inconveniences, in the opinion of most villagers, the lifestyle of pasture residency contrasts favorably with that of village residency. Life on the range is said to be more carefree, independent, and traditional. Only rarely do officials bother to make the long journey out to count livestock or insure compliance with policy regulations.

At the time of decollectivization, there were about forty households living in dispersed pastures. Then, in 1982, settlement discipline broke down when Gengden (the retired commune Party secretary) himself decided it was too inconvenient to maintain milk cows while living in the residential area. He resettled to the west of town, and many families followed his lead. Since then, the growing population has continued to expand residential settlement along the highway. Households are also increasingly building second homes out on the range. They hope to take advantage of both lifestyles, but they are also concerned about keeping a watchful eye on their neighbors' power plays for rangeland resources.

The predominant economic activity is still extensive mobile pastoralism, with defining features that best match the typology known as "distant-pasture husbandry" (see Khazanov 1994: 22). In principle, most residents are expected to live in settled residential areas while the livestock is maintained more or less on distant pastures by specialized herdsmen far from the village center. In practice, however, livestock still roam everywhere, specialists are utilized less and less, and the entire family is actively involved in a dual residence system that draws them into periodic transmigration.

Until the late 1980s, production entities known as *lianhu* banded together in clusters of two to five households to share labor, tools, and grazing land for mutual benefit. Increasingly, these informal associations have splintered into either double households (usually involving one household living in the residential center and one out on the distant pasture) or single household enterprises. *Lianhu* that remain operative typically share an explicit agreement about reciprocal obligations. For example, the pasture resident may take on all herding responsibilities during the summer season only, or for the entire year. The town resident may then pay a flat fee for this service, perhaps at the rate of 5 yuan per year for each cow or horse, and 1.5 yuan per year for each sheep or goat. Alternatively, the pasture resident may not take a wage but instead send their children into the village to board with the town resident when school is in session. The herding family also relies on those in town to take care of both group and personal business (such as production taxes), to buy commodities and deliver them on a regular basis, or to provide hospitality when they come to the residential center. Wulanaodu had about 50 *lianhu* still in operation in 1994. Informal labor swaps among neighbors also occur periodically during peak labor periods or whenever circumstances demand.

ETHNIC RELATIONS

All but 5 percent of the population of Nasihan is ethnic Mongol. In Wulanaodu village there are only twelve ethnic Han, all descendants of a single poor farmer who voluntarily migrated from Wudan in the early 1960s. He came, along with seven other Han families, when the banner government organized their transfer in response to a request by the brigade leadership for assistance in acquiring skills in intensive cultivation. Over time, all families but the one moved away. The neighboring village of Zhaoketu has a larger portion of Han residents (14 percent), and the southernmost village of Dayingzi, which borders a Han agricultural township, has a population that is 24 percent Han.

Local officials, residents, and the Han scientists at the Wulanaodu research station all assert emphatically that ethnic relations in the township are trouble free. In formal conversation, everybody repeats a standard response: "Mongol and Han are the same." Despite the claims, there is plenty of evidence that ethnicity does matter. In Chapter 1, I related how

Buhe and Longtang refused to give up a child for adoption to Han parents even though they incurred a devastating government fine for exceeding the birth quota. Repulsion also occurs at the thought of a Mongol woman being given away in marriage to a Han man. Nobody seems to mind if Mongol men take a Han wife, but residents frown upon "losing their women" to the Han, even if the couple remain in the village.

Bad feelings emerge in other domains as well. During my year of fieldwork, numerous thefts occurred that greatly concerned the entire community. Theft of livestock was especially troubling. I noticed, however, that whenever a crime was reported, residents always assumed the perpetrator was Han. This was true for all manner of thefts that occurred, with the notable exception of fence-wire. The topic of wage employment and labor migration in general provided yet another context for angry words about exploitative Han neighbors. Discussions of the research station also brought forth evidence of bad relations. During interviews, many residents took the opportunity to jump from a discussion about the work of the scientists to make gratuitous insults about Han people and Han customs in general.

But evidence of ethnic tension flared most dramatically whenever residents became intoxicated. After sufficient quantities of alcohol, it is not surprising to hear residents openly curse "Han chauvinism" and delight in recounting the glories of Chinggis Khan. Songs frequently accompany drinking, and one of the most popular folk songs recounts the story of Gada Meiren, a Mongol military hero in the early 1930s who led a rebellion to protest Han colonization in the grasslands. Furthermore, there were accounts in the village of herders who had become drunk in nearby farming townships and started public brawls. Contrary to the official rhetoric, ethnic tension does continue to characterize the region.

KIN GROUPS

Since Wulanaodu was founded, there have been three predominant kin groups organized around the surnames Yu, Han, or Wang. As part of the assimilation process over the twentieth century, various Mongol clans adopted Han surnames to facilitate their bilingual status. Mongol residents of Nasihan still maintain both a Mongolian name, which they use among family and friends, and a Chinese name, which they use for government records and official purposes. Nepotistic competition between the three dominant family lines has played a major role in village social life, influencing not only overt political decisions but also daily patterns

FIG. 5.2. Illegal expansion of household enclosure (*before*)

FIG. 5.3. Illegal expansion of household enclosure (*after*)

of social contact and resource management. Over time, each has exercised a period of ascendancy relative to other kin groups. As of this writing, the Wang family is in ascendance, and residents sharing that surname enjoy definite social privileges.

Of the original twelve founding families, six of them shared a Wang surname, including the founding authority, Gengden. Currently, the village chief, the Party secretary, and the highest ranking Party officers (including four of the five local Party positions of influence) in Wulanaodu all belong to the Wang clan. Two individuals in particular control the political strings. They are both retired Party officials who yield their clout in Party committees at the township level. They effectively maintain control over local Party discipline by controlling who may join the Party in the first place. When it comes to the allocation of local resources, these two individuals easily subvert the intentions of banner and provincial level policies to suit their own purposes. Just as Deng Xiaoping once controlled affairs in Beijing without any official title, these two men control affairs in Wulanoaodu. As the locals put it: "What they say counts" (*tamen shuole suan*).

Some of the officials within the Wang clan press their social advantages aggressively. For example, the man who presided as village chief during my period of fieldwork completed a number of conspicuous improvements to his homestead at public expense. First, he built a dam on a public thoroughfare beside his house so he could hijack a transecting stream to irrigate his private garden. He then cut down in the same road a prominent almond tree whose shade provided the community a popular place to rest from the afternoon sun. He sold the lumber and incorporated the land into his own enclosure (Figures 5.2 and 5.3). This act was especially brazen because such a tree in Nasihan is considered a "national resource" and cannot be cut without explicit permission from both the township forest bureau and banner level officials.

Residents bearing the surname Wang also enjoy substantial privileges to community resources. To offer but one example, a retired former official once established an enclosure on land near the residential center that blocked an existing public road. Many residents complained to both village and township level authorities, but no one forced the elder Wang to remove his fencing and a new road was gradually created along its perimeter. In the words of one disgruntled senior resident, "many of our village officials commit crimes for which they never face punishment or discipline because the Wangs are too powerful."

ALCOHOL

Ethnic Mongols in China are notorious for a highly cultivated drinking capacity, though scholars have tended to avoid or downplay the topic. Many Western visitors to the grasslands (both past and present) have remarked upon the prominent place of alcohol in Mongol expressions of sociability (see Gilmour 1883: 317; Montagu 1956: 87; Becker 1992: 40). Perhaps a more objective measure, the dietary survey by Chen Junshi and his colleagues (1990: 696–697) has indicated that average daily liquor consumption among all rural males in Inner Mongolia is more than double the national rural mean.[4] I can confidently report that in Nasihan, daily indulgence of alcohol goes far beyond the dictates of mere sociability. Indeed, the routine and ubiquitous display of ritualized drinking indoors and public drunkenness out of doors testifies to a deeply embedded social phenomenon that should neither be dismissed as commonplace pathology nor be disregarded as quaint local color.

As a resident guest in Nasihan, I found myself well positioned not only to observe some of the ritual dimensions of alcohol consumption, but also to experience the cultural pressures and bodily discomforts associated with liquor that sociability among rural Mongols entails. Once I started interviewing in the home, I quickly learned that refusal to drink signified to my daily hosts nothing less than a refusal to *jin reqing*—to engage them on equal and mutually respecting terms of friendship. The capacity for drink takes on symbolic representation of one's capacity for friendship, and for loyalty to a cause higher than oneself. Refusing to drink was not an option if I expected to engage the people's cooperative and interested attention (see Figure 5.4).[5] But after the liquor was opened, a battle always ensued over how little I might get away with drinking. I once found myself literally thrown into a tight headlock until I agreed to swallow one more cup, despite my protests that I was nursing an illness. The obligation to consume alcohol was definitely an arduous aspect of fieldwork, but it opened my eyes to a meaningful dimension of local behavior.

Alcohol consumption is a purposeful ritual among Nasihan herders. At a minimum, drinking may be said to play an important role in ethnic identity construction. It serves as a highly visible marker of Mongol hypermasculinity, both to fellow Mongols and to neighboring Han. Drinking, like riding horses, killing sheep, and wrestling in the grass, has become an important vestigial indicator of "otherness," a mutually crafted means of perpetuating the tough and free-spirited "barbarian" complex

FIG. 5.4. Local residents host author with food and liquor

of former days that simultaneously draws admiration and disgust from Han Chinese (see Khan 1996).[6] As I indicated earlier, evidence of ethnic tension in the region tends to flare most dramatically in association with alcohol consumption.

In addition to fortifying ethnic identity, however, alcohol consumption also plays an important role in facilitating local social integration. It does so primarily by helping to resolve a basic production problem for local residents. Low-technology animal husbandry can be viable only so long as adequate land for extensive use remains widely available and reasonably fertile. Extensive land use in turn requires dispersed settlement patterns and production units. Yet the same ecology and economy that favors small, scattered, and independent household producers simultaneously requires some degree of labor exchange and risk sharing, given the hazardous and highly unstable circumstances of life in such a marginal production environment. Peaks in labor demand occur primarily during spring shearing, summer garden cultivation, early autumn hay production, and late autumn harvest. The twin shortages of human labor and livestock fodder can only be circumvented by establishing family networks of cooperation. These circumstances have produced a community that highly values independence and competition, but vitally depends upon cooperation and alliance. Finding a common channel to assist social exchanges thus becomes a matter of great concern to all residents. Alcohol has become the favored means of resolving this tension and of pursuing differentially desired levels of integration. Indeed, for purposes of temporarily leveling social distance and economic difference, a more efficient means than alcohol could hardly be found.

Nasihan drinking is almost always a social event, from which follows consistently the idea that alcohol consumption plays an important role in the political economy. Though solitary drinking does sometimes occur, it is widely frowned upon and constitutes the only context in which residents will refer to "problem drinking." Within a village community, each man's drinking capacity quickly becomes common knowledge, and friends are expected to brag about each other's alcoholic stamina.

Alcohol consumption is a public activity, but frequent indulgence tends to be the exclusive domain of adult males. Husbands and wives often share drinks in the privacy of their home, but when other men are present, women either serve appetizers or make themselves scarce while the men get drunk. At the female gatherings I witnessed, tea was typically the beverage of choice. My wife, in her social activities, was never compelled to consume liquor in the manner I was. Still, some women do nurture their own capacity for liquor. For example, they sometimes emerge from their work at the later stages of a meal to honor a guest with numerous drinking challenges before abruptly departing again. I also observed that a few widowed women beyond the age of fifty occasionally wandered at midday through the residential center intoxicated. However, it seems to me that domestic violence constitutes the most serious alcohol-related health problem facing Mongol women.[7] In Wulanaodu, I became aware of several households in which the women and children lived in perpetual fear of male relatives who became abusive when drunk.

### The Annual Production Cycle

The annual pastoral production cycle may be considered to begin in April. By then the earth has thawed, newborn calves and lambs can be expected to survive, and herding households prepare for the spring seasonal activities. On the twentieth of the month, herders begin to shear their goats, first cutting the long, coarse hairs (*chang mao*), and then carding out the short, fine hairs (*rong mao*) used to make valuable cashmere. The whole township becomes flooded with motor vehicles carrying traders from Hongshan, Chifeng, and outlying metropolitan areas who scour the countryside to buy *rong mao* (at 80–100 yuan per half-kilo in 1994), which they can sell to urban factories at double the price. The money herders earn from *rong mao* generally constitutes a large percentage of their annual income and provides them with the cash flow necessary to recover from winter and to undertake their yearly productive investments. It also provides the source for annual spring tax payments.

The pastoral tax rates (*muyie shui*) are assessed every three years by banner officials in Wudan. Sheep and goats are taxed at 2 yuan per head, while large livestock are taxed at 5 yuan each, based upon livestock holdings reported in the previous June (when household holdings are greatest). A production tax (*chanpin shui*) is also assessed for the sale of meat, wool, and hair offtake at the rate of 4.5 yuan per sheep or goat, and 3.10 yuan per large animal, based upon livestock holdings during the previous December. Landholdings are taxed at the rate of 0.5 yuan per mu within the collective hayfield, and at the rate of 0.05 yuan per mu of enclosed land out on the range. In recent years, additional production taxes have been levied by local government authorities (without warning) on a variety of routine activities, including hay production and the transport of livestock to market. Of course, many of these taxes can be dodged or minimized by having the proper social connections. There is yet another tax assessed on household labor power (*fudan kuan*), which must be paid in lieu of providing corvée obligations to the local government. The rate is 20 yuan per year for each mature able-bodied worker. A list of total annual taxes due from each household is posted publicly on a wall in the residential center of each village. In 1994, the cumulative amount due from Wulanaodu households was 59,253 yuan.

Shearing and trading goat hair lasts until about May 10, when everyone begins to prepare their small fields for grain cultivation. First they harrow, plow, and fertilize the soil with whatever means they have at their disposal. The most destitute households simply rent their allotted field space to other more capable producers. Poor households rely upon their own labor power, scratching at the earth with a narrow iron plow that has been hitched to a cow, mule, or horse for traction. One person guides the animal along the proper furrow while another controls the plow tip. Someone else walks behind dropping the seeds. Next the soil is pulled back over the seed by someone pulling a crude scraper that drags along the ground. Finally, someone else drags the furrow with a heavy metal ball that packs the soil down over the seed to prevent exposure to wind and birds (see Figures 5.5 and 5.6). In most cases, the only fertilizer that residents apply to their fields comes in the form of dried sheep and goat manure. Cow dung is deliberately set aside (and also scavenged from

FIG. 5.5 (above). Horse drawing plow tip across field.

FIG. 5.6 (left). Woman dragging field with weight to pack down soil

FIG. 5.7. Old tractor from collective era

the open range) to be burned as household fuel. Only a few residents bother to purchase the expensive chemical fertilizers from urban markets.

Households with better financial resources are willing to pay cash for more elaborate and intensive field preparation. In Wulanaodu, there is a single 1950s-era tractor that once belonged to the commune but passed to the former commune mechanic upon decollectivization (see Figure 5.7). This resident now charges each household between 15 and 20 yuan per mu of land for his field preparation services. I noticed that he tended to charge higher rates to those who had the least social status. In 1994, about forty different households in Wulanaodu hired him to harrow and till their fields. These clients all became quite anxious about planting their crops on time when heavy rains fell for three consecutive days and created a backlog in the tractor's service schedule. The mechanic admitted to me that he prioritized his workload according to the ability of each client to pay—lower social status again tends to amplify production handicaps.

Also in early May the vegetation turns green, almost overnight. As the grass begins to grow, those households who can afford the expense erect new fence-wire for the coming year and make repairs on existing enclosures. The busy pace of the spring season creates acute labor shortages and motivates some residents to organize small-scale labor swaps (*huan gongzu*). Typically, several households who share either close kinship or geographic relationships with the host of a work party will each send a single representative to contribute a day of labor. Labor swaps are usually arranged only for large projects that need to be completed quickly, such as building a house, cultivating a field, or harvesting produce. Those who contribute labor in this way then enjoy the reciprocity of gaining extra labor power whenever they might need it. The largest labor swap I witnessed involved ten adults from nine different households who gathered to help till and sow a large field for (illegal) cultivation. The host of the work party borrowed the labor of helpful neighbors even as his own two sons left the scene to participate in a different labor swap with residents of another village.

June is the season for shearing sheep and marketing wool (*yang mao*). In 1994, wool was traded at a price of 4 or 5 yuan per half-kilo. Once sheared, both sheep and goats are then treated for disease with special chemical sprays and injections. These animals are also culled and sold for meat throughout the spring and early summer because they mate in the late autumn. Only a few males are kept in each flock (for up to six years) to serve as stud; the others are castrated as early as a month after birth. Mutton sold for about 1.5 yuan per half-kilo of live weight in 1994. An average young sheep might weigh about 50 kilograms and bring its owner roughly 150 yuan. The trade in sheep and goats is much less restrictive than the trade in cattle. Many herders sold their surplus sheep and goats to local individuals who simply marched them over sand hills and along public roads on a two- to three-day journey into the processing factories at Wudan for a slender margin of profit. Many residents keep back a portion of their wool to make into felt. No local resident has the skills to do this, however, so itinerant artisans (from Han villages) come and sell their services at this time of year (see Figure 5.8).

After the flocks are culled, sheared, and vaccinated, gardening occupies the energies of most households, especially the chores of weeding and irrigation (the latter for those who can afford the equipment). At this time, some of the more diversified households begin to stock their ponds with fish. They later catch the fish by throwing chemical pellets into the water, which anesthetize the fish and bring them floating lifeless to the surface where they can be easily retrieved. Otherwise, the herders allow itinerant Han traders with more experience to gather the fish at a set

FIG. 5.8. Artisan cleaning wool for felt

price (such as 100 yuan per basket) and sell them in urban markets at a huge profit.

Early summer is also marked as a popular time to host and attend *oboo* festivals and *nadaam* celebrations (see Figures 5.9 and 5.10). An *oboo hui* is a private social event marked by games, feasting, and sometimes religious rituals focused upon an outdoor altar or pile of stones that

FIG. 5.9. Local *oboo*

FIG. 5.10. Wrestling at local *nadaam* festival

FIG. 5.11. Residents let their animals wander through the residential center

is usually sponsored by a single household who uses the occasion to enhance its influence and expand its network of alliances in the region. *Nadaam* is a public periodic social event focused upon competition in a variety of "manly" games such as wrestling, horsemanship, archery, and long-distance running. Residents assert that this ceremonial display of sport and military prowess dates back to an era preceding Chinggis Khan.

Throughout the summer, herders get on with the daily chores of managing their animals. Some of the skilled households are extremely attentive—leading their herds and flocks separately to preferred grazing locations, milking the females on a regular schedule, ensuring a proper water supply through the day, and securing their safe return to a protective pen every evening. Many others, however, are much less involved and allow their animals to wander on their own accord both day and night through the residential center and out on the range (Figure 5.11).

By late August, residents begin to prepare for autumn harvest and winter survival. Women peel vegetables (such as turnips, eggplant, and tomatoes) into slices, then hang them in the sun to dry so the strips can be boxed and stored for consumption during the cold season. Other vegetables (like cabbage) will be boiled down, salted, and stored in large open bins to be consumed with potato noodles and small portions of shredded meat. No one preserves food through a canning process. This month is also a popular time for marketing timber. Upon decollectivization, tenure rights to some forested areas and tree stands were granted to private households. Once the trees reach a certain girth, households obtain permission from the local branch of the Ministry of Forestry and then sell a limited number of them each year to outside buyers. A mature poplar tree, for example, can sell for about 50 yuan.

September is probably the busiest month of the year, as each household begins the huge project of cutting and storing the hay that will keep their livestock alive through the long winter months. The summer rains have ended and the grass must be cut, dried, raked, tied into bundles, stored, and then eventually transported to distant pasture homesteads. Cutting the tall grass in the hot sun amid swarms of mosquitoes, flies, and hornets is the most difficult phase of the entire process. There are no mechanical mowers in this poor region of China, so everyone uses the same simple technology—a long wooden handle with a steel blade that must be swung in a wide arc along the surface of the ground.

Although the work is exhausting and continues from sunrise to sunset for several weeks, there is a certain ceremonial and communal quality to it, as each household descends upon the same reserve hayfield at the same time to work their allotted strips of land. They each set up a temporary campsite in the hayfield as a shaded place for rest, meals, and socialization. There is plenty of time for sipping tea and chatting since the sickle blade must be resharpened against a whetstone after cutting each row. All able-bodied family members participate, and even young children come to play at the campsite and help out with some of the easier chores (see Figure 5.12). Despite the camaraderie, the dominant mood seems to be a shared sense of drudgery and resignation rather than overt celebration. For example, the cartloads of people and equipment move silently through the village in long somber lines as they begin and end each day of grass cutting. Even when visitation occurs at a campsite, people tend to stay focused on the work at hand and do not launch into jovial stories or festive conversation. The only hint of celebration derives from the fact that households begin to slaughter their sheep and goats for domestic consumption, apparently because of the need to consume more protein at this moment of peak labor demand.

Once the hay is properly stored for winter, agricultural harvests follow

FIG. 5.12. Household enjoys campsite while cutting hay in the reserve pasture

in the month of October. First, the soybean and then the corn is cut from the field and laid out to dry. Later the beans are threshed and the cornstalks cut and stacked into piles for livestock feed. Much of this produce is saved for domestic use, but some households do sell their surplus crops to other neighbors or itinerant traders. Once the crops are managed, and the livestock have been fattened somewhat, herders begin the process of culling their cows for market.

Government contracts for household production quotas were discontinued years ago, so residents are now free to sell as much or as little meat as they want to whoever is willing to buy. Most households sell their select animals to private entrepreneurs from urban areas who drive large trucks into the village and negotiate prices on the spot. A big steer may yield 200 kilograms of meat, and at a rate of 7 yuan per kilo (no matter the age of the animal), a herder can pocket about 1,400 yuan per head. Other individuals with more financial resources prefer to rent their own equipment and transport their animals directly to regional stockyards for a considerably better price and less guesswork. The stockyard I visited paid 3 yuan for each kilo of live weight measured on a proper scale, which might bump the market price of the same steer up to 1,600 or

FIG. 5.13. Resident setting up strategic windbreak in preparation for winter

1,700 yuan. Nasihan herders want to sell at the most advantageous price, of course, but they generally are not inclined to sell more animals just because of a good market price. Unless they are in special financial distress, they prefer to sell only the number of animals they cannot expect to keep alive on hay supplies through the winter. Most households still consider a growing herd to be the easiest and quickest route to prosperity. In the words of one highly successful resident, "If I sell at a high price, I just have to buy them back again at the same high price."

During the month of November, households continue to make marketing decisions about their herds. They also make final preparations for winter. This involves weatherproofing the homestead with strategic windbreaks that are handwoven from willow branches, fortifying the walls of the house with new layers of mud, adjusting the size and sturdiness of animal pens, and collecting as much manure and firewood as possible (see Figures 5.13–15). They also stockpile supplies of grain and other com-

FIG. 5.14. Resident preparing for winter by mending cracks in the wall of his house

FIG. 5.15. Stockpiling manure and wood for the winter

FIG. 5.16. Scavenging tree branches from artificial forest for winter fuel

modities to avoid winter travel as much as possible. Once these preparations are complete, many households undertake a short season of hunting. For about two weeks, the sound of distant rifle fire may be heard
from the village center, as scarce wildlife (pheasant, duck, and rabbit) run
for cover in the artificial forest areas. More exotic animals (such as
wolves, boar, and gazelle) have long been eradicated by work brigades
during the collective era.

The hostile winter months of December, January, and February are
characterized primarily by social activities. This is the season for wedding
festivities and for extended visitation with family members. Migrant laborers return home for the winter, some households travel away from the
village for weeks at a time, while others entertain house guests for long
periods. Neighbors often drop by to play a friendly game of chess, to
trade reading materials, or to share a bottle of liquor. There is very little
activity outdoors, except for the daily survival chores of drawing water,
scrounging for dung and firewood, and distributing hay rations to the
livestock. In late winter when supplies run low, residents use long poles
to scrape off high branches in the artificial forests and fell entire trees for
fuel (see Figure 5.16). They remove tree bark from logs and give it to the

animals to gnaw on. The season is also marked by a high frequency of cold weather accidents, as people work and travel in bitter temperatures while under the influence of alcohol.

By March, residents are eager for a change of pace and dismantle their winter fortifications around the home. Hay is now a precious commodity, and households who have some to spare can sell it to desperate neighbors at inflated prices. Nearly all the village livestock appear emaciated from winter food stress. Out on the range there is nothing left to eat—even many of the woody shrubs have been chewed to the roots. This is also the season when contractors arrive in the township to entice as many residents as possible into becoming migrant workers. The offers are tempting to young workers, despite the widely circulated tales of misfortune that have befallen previous migrants. Life-changing decisions must be made virtually on the spot, as the contractors come and go without notice and without patience. By the end of the month, general household production strategies have been set for the coming year, and the cycle begins again.

From this brief review it becomes apparent that Nasihan has been buffeted by the dominant currents of modern Chinese history. Lattimore (1934) once dubbed the region of Northeast China the world's geopolitical "cradle of conflict," in anticipation of World War II. After living through the turbulent successive eras of Manchu colonialism, Nationalist indifference, Japanese occupation, indigenous warlordism, Communist land reform, Cold War tension, and the violent political activism leading first to collectivization and then to decollectivization, local residents and their offspring must now confront the comprehensive and disorienting challenges of economic globalization.

Despite the waves of change, the community has clung to a traditional (i.e., nineteenth-century) pastoral identity, basically burrowing its head in the deepening sand until political storms pass and familiar norms can return. Without endorsing the condescension of an English missionary who toured the Keerqin desert at the turn of the century, I find it useful to cite one of his subjective remarks. It speaks to the overwhelming sense of isolation and marginality that he sensed from the lay of the land and the pace of social life. Referring to the inhabitants of eastern Wengniute, he wrote: "The world may grow old, kingdoms rise and fall, social and political questions may be throbbing in the lives of civilized nations, but these simple, ignorant denizens of the desert live their lonely life in quietude, knowing nothing and caring less for the world that lies beyond them" (Hedley 1910: 238).

Until the reforms of the 1980s, Nasihan was a place that was badly bruised, yet fundamentally unmoved, by historical turbulence. But changes associated with the reform era have accelerated the pace of life and the pace of social transformation in primordial ways, and community members increasingly sense themselves approaching a crossroads in their personal and collective identity that they can no longer ignore. It is unfortunate that a moment of such opportunity is darkened by the shadows of suspicion that remain from a long legacy of economic exploitation and political fear.

# SIX

# Enclosure and Changes in
# Physical Landscape

In recent years, the Chinese central and provincial-level governments have formulated grassland policies to address human dimensions of resource management. Ironically, interventions imposed from outside local communities have, in many respects, only made matters worse for much of the native population. Whatever the combination of physical, biological, and social forces that may have converted grassland to sand historically, in Nasihan it has been the grazing management system itself that has magnified wind and soil erosion processes leading to desert expansion over the last two decades.

Since decollectivization, the favored policy to control the process of ecosystem decline has been a movement toward pasture enclosure through the construction of wire fences. Once implemented at the village level and mediated by local social realities, however, this policy actually accelerates the very chain of decline it intended to bring under control. That assertion may well sound counterintuitive to anyone who has toured the grasslands, where even casual observation confirms a consistent structure to the landscape: enclosed land appears green from light grass cover, whereas land outside the fences appears patchy at best, marked by semifixed and mobile dunes. Given such stark contrast, it is easy to imagine enclosures to be a tidy solution to the problems of resource management. Perhaps it is just such superficial tourism (supported by political and cultural motivations to misperceive local realities) that accounts for the longevity of this erroneous but popular view among Chinese policy makers and scientists. Despite appearances, fences are not saving the grasslands. If one stays long enough to observe local behaviors, a very different perspective emerges.

As privately enclosed land area has steadily increased since 1980, local elites have advantageously manipulated stocking ratios in such a way as to intensify grazing pressures on tracts of land most immediately vulnerable to erosion processes, and they do so at the most vulnerable phase in the vegetation growth cycle. In essence, the modest greenery of private enclosures is purchased at the expense of the larger regional ecosystem. This chapter will first review the intellectual origins of the enclosure movement and post-reform policy initiatives, and then provide the evidence to support the argument against the enclosure movement.

## Indigenous Origins

The practice of enclosing portions of rangeland to protect it temporarily from livestock grazing pressure was apparently indigenous to Inner Mongolia. The Mongols have a native term (rendered in Chinese as *kulun*) that refers to naturally occurring dune-enclosed meadows or isolated plots of grass where wind, water, and soil conditions create favorable pasturage. According to local residents, the minor use of fence-wire to enclose small natural meadows first began as a method to preserve the *kulun* patch structure.

During the collective era of the 1950s, when the rural population increased and Mongol herders organized into fixed communities, brigade leaders in many locations first decided to enclose and set aside the most fertile meadows for communal winter hay production. According to Hu Mingge (1994), a senior director of the Animal Husbandry Bureau in Chifeng City, the practice first began in western Inner Mongolia in Yikezhao League. Before that, the term "enclosure" was used in reference to warming pens for livestock. In 1963, Wulanaodu itself enclosed a field of roughly sixty hectares for use as such a collective reserve.[1] While the sudden decision to utilize fence-wire was indeed made by leaders at a local level, those decisions must be understood within the wider context of national collectivization. In grassland areas where land degradation was already advanced, the collective era promoted some aggressive experimentation in land rehabilitation.

One of the most influential experimental sites (from as early as 1964) was the Wushenzhao (Uxinju) Commune in Wushen banner. The people of Wushenzhao first devised and tested the "four-in-one" *kulun* (water-grass-forest-cereal), which may well have been the national prototype for more intensive household land use policies promoted after the economic reforms of the 1980s. The four-in-one method specified a systematic use of enclosed land, devoting 5 percent of total acreage to growing selected grasses and cereal fodders, 5 percent to timber resources, 30 percent to improved hay, and 60 percent to natural pasture. The whole enclosed area was protected from summer-fall grazing.

By 1978, the Department of Desert Research in Lanzhou excitedly proclaimed such enclosures an effective means to control erosion and to reconstruct damaged grassland. With high-profile endorsements, they encouraged similar initiatives:

The Uxinju People's Commune . . . has shown that building grass kulums [sic] is an effective measure for preventing damage from drifting sand; for protecting, managing and rationally utilizing pastures; and for building sustained high-yield fodder bases. . . . Kulums [sic] of this type have come to play the major role in reclaiming and improving the grasslands, and are increasing at a rapid rate. (DDR 1982: 16–18)

From the beginning, enclosure practices were intended to keep animals off specified plots of land, allowing them to forage inside only on the stubble that remained after fall mowing (ibid.: 51). With this method, hay production was reported to increase by nineteenfold, but such impressive results accrued after fifteen years of intensive care involving both artificial seeding and perpetual summer fallowing (ibid.: 20).

Pioneering experiments with grassland enclosures during the collective era seemed very promising to grassland officials. Yet these experiments operated under different technical and social dynamics than have existed on the range since decollectivization. Before the reforms, there was no zero-sum dynamic in operation between public and private land, because all land was owned collectively. Therefore, enclosures could actually function as an instrument of controlled rotational grazing under organized management. Also, the size of enclosed territory remained tiny relative to open rangeland. Today, however, as the rest of this chapter will make clear, the situation is fundamentally different, so that enclosures no longer play the same role in rationally distributing local grazing pressures for the mutual advantage of all residents. The early Chinese optimism for modest rangeland enclosure practices was based upon laboratory conditions that totally discounted factors such as scale and disruptive social influences that would significantly affect future policy implementation.

## Intervening Influences

The utilization of grassland enclosures as a rangeland management technique in IMAR may have indigenous origins, but the collective era never introduced anything that compares, either in scale or operation, with contemporary enclosure practices. A modest indigenous idea quickly spiraled beyond the control of local brigade leaders and individual herders. Intervening outside influences, both domestic and international, have played a role in accelerating and proliferating the household use of fence-wire in pastoral Inner Mongolia since the early 1980s. Chifeng City prefecture has been a focal point for several government-sponsored pasture intensification programs, primarily because of its high livestock-to-forage ratio (Hu 1994).

## DOMESTIC FORCES

One set of actors who have been highly influential in proliferating household enclosures are the various political reformers within China itself. Officials and scientists at all levels of government have promoted the privatization and parcelization of rangeland resources. Anxious to prevent a "tragedy of the commons" scenario from unfolding on national rangelands, the Chinese reformers have asserted that changing demographic realities require bold policy initiatives to reorganize grassland production. Their first move was to introduce the pastoral "double contract" household responsibility system of management (*xu cao shuang cheng bao*), whereby local production brigades distributed land use rights (in 1984) as well as animals (in 1981) among independent herding families.[2]

Privatization of rangeland resources was supposed to be just the first step in a long series of adjustments intended to "rationalize" the animal husbandry sector. Li Yutang, Director of the Grassland Division of the Ministry of Agriculture in Beijing, has outlined the basic reform strategy: first, distribute animals to private households; second, distribute grazing lands; third, assign carrying capacities for each plot of land; and finally, implement incentives and sanctions to enforce a sustainable balance between animals and vegetation at the household level (NRC 1990: 33; Li Yutang 1992). Prominent scientific publications have likewise endorsed a similar formula to attain the desired "rational management system" (see Huang 1989; Jiang Su 1989; Zhang Xinshi 1989). Interviews with senior county-level officials in Wengniute confirm that existing policies intend to eliminate the practice of extensive grazing. The direction of Chinese animal husbandry policy in Chifeng City prefecture and beyond is toward large-scale, intensively managed feedlots (Guo 1993).

From the beginning of reform, policy makers have assumed that private enclosures would force independent households to confront the extreme contradictions between forage demand and forage availability among their separate herds. The logic of that policy requires animals to be contained within a bounded territory. Furthermore, household managers were expected to gradually substitute labor and capital for increasingly scarce fertile land, and thus a traditional, "rely upon heaven," extensive grazing regime was expected eventually to yield to a conservationist, "scientific," intensive regime made possible through technical inputs and motivation born of private ownership.

The following statements are representative of conventional wisdom on the subject: "Fencing the degraded grasslands to exclude grazing animals for some periods is one of the most economical and effective ways to recover grassland productivity" (Chang, Cai, and Xu 1990: 461). "Grass yields have been doubled or quadrupled merely by fencing the original pasture. . . . Productivity will be enhanced if pasture improvement and scientific management are put into practice" (Zhou Li 1990: 44). Similar praise can be found in virtually any Chinese account of grassland problems and solutions (see Hu 1990: 210–211; Wang, Wang, and Zhang 1993: 31; Cao et al. 1984; Jiang Fengqi 1984).[3] Fences are also widely recommended on principles of social cohesion. They are said to empower private citizens to work in partnership with the government to rehabilitate national rangelands (see Xinhua 1985; *China Daily* 1988; Ba 1993).

A central problem with the scientific literature on enclosures is the continued assumption that grassland experiments conducted under controlled conditions can be broadly replicated in poor pastoral villages. Most Chinese rangeland scientists simply do not consider the social context of implementing their enclosure recommendations. They typically address the issue of enclosures as if the only relevant question were whether long-term fallowing restores the net primary productivity of rangeland vegetation. Their studies show (unsurprisingly) that it does. All well and good, but the studies do not indicate where and how the livestock should graze in the meantime, or how independent households should coordinate their rotational grazing, or how the majority of herding households might acquire the equipment to turn dunes into productive meadows. Yet this is the crux of the problem. The benefits reported in scientific enclosure experiments derive from displacing local livestock and human populations on a long-term basis, but that practice cannot be replicated on a large scale once the majority of private households begin to enclose.

Thus, despite visionary objectives, the policies have not produced the intended effects. One simple reason is that the vast majority of households have been left to their own devices to cover the high costs of capital investment. But more complicated social factors are also involved, with the result that most animals still do not graze within private enclosures, and more sustainable land use practices have not evolved.

## INTERNATIONAL FORCES

International organizations and their agents constitute another set of influential actors who have directly and indirectly encouraged the proliferation of household enclosures, especially in Chifeng City prefecture and throughout Wengniute banner. For example, at the beginning of the reform era, the United Nations Development Program and the Chinese Bureau of Animal Husbandry jointly financed and collaborated on a project demonstration center devoted to grassland development. It was located along the banks of the Xilamulun River in eastern Wengniute (called the "Wengniute Ranch"). From 1979 to 1983, the project attempted to test the viability of mechanized fodder production and to demonstrate recommended methods of livestock management to local herders. One specific priority was to increase and improve the use of fence-wire, and to teach principles of rotational grazing within enclosures. Toward that end, the project erected some thirty-seven kilometers of permanent boundary fencing, as well as additional fencing to close off dune areas for stabilization and to secure other sites for artificial seeding. In addition, foreign technicians demonstrated various bracing techniques necessary to improve fence-wire tautness (Kernick 1980: 8). The ranch originally operated these enclosures under collective management, but the assets and the techniques passed to private households upon decollectivization.

Another direct route of international influence on household enclosure proliferation in Wengniute occurred via the International Fund for Agricultural Development (IFAD). IFAD, a United Nations agency based in Rome, supported the first multilateral assistance loan to China, providing a total of U.S.$35.4 million interest-free as part of a project whose total cost ran about U.S.$64.6 million. From 1981 to 1988, the Northern Pasture and Livestock Development Project (NCPLDP) attempted to introduce and demonstrate improved technologies and practices in grassland forage and livestock production. It provided funds for a wide range of investments in fodder development, dune fixation, seeding, irrigation, breed improvement, dairy development, and marketing and transport operations. It was supervised by the UNDP and affected eight counties: four in IMAR, three in Heilongjiang province, and one in Hebei province.

Four of the eight counties (all of the IMAR sites) were located in Chifeng City prefecture, including one in Wengniute itself.

IFAD fully intended to make a significant impact on local production methods, as the 1981 annual report explained: "Although limited to a small fraction of the country's pasture resources, the project is expected to have a substantial demonstration effect and the possibility of extensive replicability" (IFAD 1981: 34). The largest investment category for the NCPLDP project was that of "pasture development," which IFAD explicitly defined as the introduction of improved grasses, seeding, fencing and shelter belts. The fundamental development strategy assumed that "properly designed, located and installed fences would contribute to fodder production and improved rangeland management" (IFAD 1989: 4). Originally, project funds were utilized for strategic collective enclosures operating under collective management. Over time, NPLDP project implementation evolved to keep pace with rural decollectivization, so that funds were increasingly distributed to "specialized households." As land was contracted out to individual households, IFAD loans were also allocated to households over a five- to ten-year repayment period, with an annual interest rate of 1.5 percent. By 1988, a total of about 20,000 loans had been disbursed, primarily to independent or small groups of households (IFAD 1989: 10).

How influential was the project in relation to household fencing? It is hard to determine conclusively. The enormous sweep of reforms opening rural China to trade, travel, and capital flow during the same time period make it virtually impossible to isolate or measure the direct effects of the IFAD project. Still, according to domestic Chinese news reports circulated at the end of the project term, the original intention to "transform pastoral areas" came to some fruition, as manifest by the following details: IMAR project sites expanded grassland fodder production by 39,000 hectares and expanded timber resources by 11,000 hectares. The project also sent many individuals abroad to learn advanced livestock management techniques (Xinhua 1988). According to internal IFAD reports, a total of about 189,000 hectares were fenced in IMAR over the course of the project (IFAD 1989: 16).

Brown and Longworth (1992: 1667) likewise assume that the injection of large sums of money over eight years made the IFAD project a significant source of new investment in Chifeng City prefecture. With some surprise, however, they report that the project apparently had little impact on the rate of household investment in fencing or on household incomes when compared to nonproject townships (ibid.: 1669; see also Longworth and Williamson 1993: 317). A number of explanations for this interpretation are possible. The authors themselves explore several, including the possibility that some of the loan income was diverted for other uses, or that loan funds were made available to nonproject townships to spread the benefits over a broader area (Longworth and Williamson 1993: 317–318). I would make two additional points. First, the nonproject townships used as control data to assess IFAD influence may well have enjoyed their own sources of fence-wire acquisition. Wulanaodu, for example, acquired a substantial supply by virtue of its status as a "model collective." Second, the opposite scenario must be considered. Given the scarcity of fence-wire in the region under normal circumstances, it is possible that even households who did acquire loans with the intention to purchase wire might not have been able to find it during the specific years when control data were gathered, thus diluting the appearance of IFAD project influence.

Without more knowledge of such intervening variables, it is difficult to assess the full impact of IFAD or other development assistance programs on the proliferation of enclosures throughout Chifeng or Wengniute. Yet the probability of their influence cannot be dismissed. In any case, the many years of IFAD and UNDP enclosure rhetoric delivered to banner officials and private households over a critical phase of policy formulation could not have failed to accelerate the domestic social momentum favoring rangeland parcelization, at least in these few project sites, if not in all of North China.

## Policy Impact

The distribution of privately managed rangeland and the subsequent promotion of household enclosures resulted in the creation of distinctly different categories of land throughout Nasihan. Within residential districts, there now exists four categories of land: (1) privately enclosed land, usually utilized for hay production, reserve grazing, gardening, or tree nurseries; (2) privately contracted but unenclosed land; (3) unenclosed collective land (both [2] and [3] exist primarily as narrow sections of compacted turf that function as thoroughfares); and (4) enclosed collective land, held in reserve from June to mid-October as fields for hay and fodder production to carry the community through lean months of winter and spring. User rights to the hayfield were originally distributed in 1982. Mowing strips were assigned to each household by lottery, with the size determined according to a formula based 40 percent upon house-

FIG. 6.1. "Artificial pasture"

hold livestock and 60 percent upon male population within the household. In addition, each household received discretionary use rights to a strip of land about 7 meters wide and 300 meters long on which to grow corn, soybeans, or additional hay.

Out on the range, removed quite a distance from residential districts, only two principal varieties of land exist—enclosed and unenclosed. Theoretically, each hectare of rangeland has been equitably distributed among all households in allotments ranging roughly from forty-seven to sixty-seven hectares in size, depending upon the quality of pasture conditions and the remoteness of pasture from the village residential center. Over the years, barbed wire of various quality has come to define vast stretches of privately enclosed land. These appear as disconnected islands of green vegetation surrounded by a pale ocean of semifixed and mobile dune sand. Within the rangeland enclosures, some areas may be more intensively nurtured for fodder production than others. Such areas are designated "artificial pasture" (*rengong caodi*), in distinction to the term "semi-artificial pasture" (*ban rengong caodi*), which applies to any fenced land (see Figure 6.1). Territory that remains outside the fences constitute what is still the largest category of land in all pastoral China: unenclosed land utilized as public range.

The impact of enclosures on local land use and grazing management strategies has been dramatic. Contrary to the central government's intention, residents graze their animals as sparingly as possible on their own enclosed land. This behavior allows many local households to maintain larger herds than would otherwise be possible, but their increase comes at the expense of their neighbors. Since decollectivization, those households who could actually afford enough costly wire to protect their pasture allotments have faced no pressure whatsoever to alter their traditional grazing habits, and so keep livestock outside enclosures as long as forage is available on the wide unenclosed range. They essentially pick clean the grass of those too poor to fence, saving their own for hay production or emergency grazing during winter and spring. Indeed, it is not uncommon for households with large enclosures to graze greedily all year long outside their own fences. More often, cows and valuable transport animals (horses, donkeys, camels) are stall-fed throughout winter and spring, while sheep and goats are enclosed anywhere from one to the full six months of cold weather. But during the six months of summer and fall, only select transport animals ever see the inside of a private enclosure.

Those residents who managed to acquire fencing early—either through direct purchase, the decollectivization process, or through social connections that provided access to limited bank loans—have enjoyed a tremendous advantage in local competition for present and future grassland resources. As one might expect, the most immediately productive tracts of land were coveted and enclosed first, regardless of proprietary contracts held by others. Furthermore, those with financial leverage have enclosed far more than their allotted share of rangeland, essentially squatting until the day that someone dares to push them back on rightful boundaries. Over time, more and more households have gained the means to claim some portion of rangeland, but they in turn have adopted the same exploitative grazing strategies.

Both the ambiguity of boundaries as stipulated in land tenure contracts and displays of nepotism on the part of local officials have been instrumental in allowing this to happen. But more entrenched factors are also at work. Limited access to rural credit is one problem. Since the enclosure policy was launched, only eight different households in Wulanaodu have managed to secure loans significant enough (in excess of 2,000 yuan) to finance production investments. These went to village elites—members with clout in the Wang kinship network, the Communist Party, or both. But perhaps the most important explanation for exploitative land use practices comes from the fact that such opportunism is actually encouraged, rather than curtailed, by grassland institutions and policy statements.

After the distribution of land use rights to households in 1984,[4] policy statements circulating among township and village-level officials stipulated that land must be well managed as a condition of tenure. Otherwise, pasture resources might be confiscated and given to households more capable of using it productively. Good management is defined in those documents as a two-step process involving, first, "protection" (*baohu*), which means surrounding property with fence-wire, and, second, "construction" (*jianli*), which means developing productive capacity by planting grass and trees (Wengniuteqi renmin zhengfu 1984). Within Nasihan villages, this condition has been construed as outright license for the illicit possession of property—whoever has the leverage to enclose land by definition becomes its rightful caretaker. This policy perpetuates the Han cultural notion that any uncultivated land is barren and waste, and that any resident who does not participate in the "construction" of rangeland does not deserve a place on the landscape. If land is not fenced and intensively utilized, others may rightfully confiscate it. In practical terms, the enclosure policy thus dictates that those residents who are either too poor to buy fence-wire—or simply uninterested in restructuring the local environment along the lines prescribed by central government authorities—will be eliminated as unfit for modern China, an evolutionary dead end unworthy of legal protection.

Policy has thereby become an unintended source of economic exploitation and chaotic grazing practices. This is the real context in which influential scholars and policy makers in Beijing, apparently oblivious to local conditions, repeat their calls for "rational" land use that extol coercive measures (see Chifengshi caoyuan jianlisuo 1994: 169). Under prevailing conditions, their sanctions would not penalize exploitative elites or despotic landlords (*eba*, as residents call them), but would instead punish the majority of poor households who are losers in the race to control community resources.

Beyond the problematic implications of such land use for social justice, the enclosure movement has both perpetuated and aggravated preexisting land tenure insecurities. Land tenure institutions have changed far too often over the last fifty years for these settled herders to see the sense of investing much energy or capital into pasture reconstruction. Residents of Wulanaodu simply laugh when asked how long they expect to enjoy utilization rights to the land over which they now assume control. During my formal interviews in Wulanaodu, at least 52 percent of household managers openly doubted the long-term stability of current tenure relationships. The free-for-all atmosphere created by unregulated enclosure practices has only fed this insecurity for all.

In addition to the problem of illicit enclosures on the range, land tenure insecurities are further aggravated by changes in policy regarding access to the collective hayfields that are reserved for winter fodder. For example, mowing rights were redistributed in 1987 to accommodate the population growth and provide for newly created households. For that distribution, strip size was based solely upon family population because livestock holdings had already become too unequal. In 1993, local authorities in some Nasihan villages decided to raise revenues by charging residents substantial new fees for their mowing rights, which must be paid in advance. People are free to refuse the new tax, but they would thereby forfeit their rights to harvest precious winter fodder. These changes occur without warning or any guarantees about future rates.

## Catalyst of Land Degradation

As enclosures expand, grazing pressure and erosion intensify on the public range, while the poorest residents bear the brunt of ecosystem decline. Seeking to prevent a "tragedy of the commons" scenario, enclosure policies in Wulanaodu have actually precipitated one. The data are clear.

Table 6.1 reports the increase of privately enclosed land in Wulanaodu since 1980. (Some land was enclosed prior to redistribution in anticipation of imminent reform.) Table 6.2 reports yearly "sheep equivalent units" (SEU, see below) in the village for the summer and fall months from 1984 to 1993. It also indicates yearly stocking ratios on unenclosed and enclosed public range during the critical vegetation growth season from late spring to late fall. (A stocking ratio is the measure of livestock grazing pressure per unit of land, usually reported as SEU/hectare.) Table 6.3 likewise reports yearly sheep equivalent units for the winter and spring months from 1988 to 1994 (the only years Wulanaodu officials collected such data). It also tracks seasonal stocking ratios within household enclosures. The reason for the large difference between summer and winter livestock figures is that large numbers of animals are sold or slaughtered in the later months of the year, after they are fattened somewhat on fall harvest but before the cold season begins.

The data reveal several important trends.[5] First, stocking ratios on the

TABLE 6.1

*Yearly and Cumulative Land Enclosure Figures, Wulanaodu*

(hectares)

| Year | Yearly total | Cumulative total | % of range | Average household enclosure size | Households with enclosed land |
|---|---|---|---|---|---|
| 1980 | 471.57 | 471.57 | 7 | 11.4 | 8 |
| 1981 | 113.72 | 585.29 | 9 | 17.1 | 12 |
| 1982 | 667.93 | 1,253.22 | 19 | 39.6 | 22 |
| 1983 | 183.76 | 1,436.98 | 22 | 33.0 | 32 |
| 1984 | 270.67 | 1,707.65 | 26 | 30.9 | 43 |
| 1985 | 223.11 | 1,930.76 | 29 | 30.4 | 51 |
| 1986 | 407.87 | 2,338.64 | 35 | 31.6 | 62 |
| 1987 | 394.00 | 2,732.63 | 41 | 32.2 | 73 |
| 1988 | 368.52 | 3,101.15 | 46 | 34.9 | 78 |
| 1989 | 244.46 | 3,345.61 | 50 | 33.4 | 88 |
| 1990 | 318.03 | 3,663.63 | 55 | 34.9 | 94 |
| 1991 | 82.51 | 3,746.14 | 56 | 34.0 | 99 |
| 1992 | 50.89 | 3,797.03 | 57 | 32.8 | 104 |
| 1993 | 29.41 | 3,826.45 | 57 | 32.2 | 107 |

SOURCES: Interview data; Wulanaodu Gacca official document 1984–1993.

TABLE 6.2

*Summer-Fall Sheep Equivalent Units and Rangeland Stocking Ratios, Wulanaodu*

(SEU/ha)

| Year | SEUs | Unenclosed ratio | Enclosed ratio |
|---|---|---|---|
| 1984 | 16,531.8 | 3.307 | 0.093 |
| 1985 | 15,111.6 | 3.151 | 0.114 |
| 1986 | 16,252.0 | 3.700 | 0.116 |
| 1987 | 15,890.6 | 3.969 | 0.112 |
| 1988 | 19,290.3 | 5.292 | 0.148 |
| 1989 | 20,272.2 | 5.961 | 0.154 |
| 1990 | 14,865.1 | 4.769 | 0.161 |
| 1991 | 14,682.1 | 4.827 | 0.169 |
| 1992 | 17,203.7 | 5.773 | 0.181 |
| 1993 | 16,204.8 | 5.483 | 0.178 |

SOURCES: Interview data; Wulanaodu Gacca official document 1984–1993.

unenclosed range have climbed well beyond recommended levels year after year. Officially sanctioned stocking ratios on degraded range in Wengniute have an upper threshold of 2.15 SEU/hectare. Banner-level officials estimate that the range in its current condition can produce on average only about 925 kilograms of forage per hectare (or 61.71 kg/mu) over the

TABLE 6.3

*Winter-Spring Sheep Equivalent Units and Range*
*Stocking Ratios, Wulanaodu*

(SEU/ha)

| Year | SEUs | Enclosed ratio | | Unenclosed ratio | |
|------|------|--------|--------|--------|--------|
|      |      | Winter | Spring | Winter | Spring |
| 1988 | 17,361.8 | 2.876 | 2.327 | 2.415 | 2.793 |
| 1989 | 15,577.7 | 2.606 | 2.140 | 2.361 | 2.492 |
| 1990 | 11,814.1 | 1.855 | 1.422 | 1.690 | 2.110 |
| 1991 | 13,171.3 | 2.092 | 1.633 | 1.855 | 2.322 |
| 1992 | 14,243.3 | 2.381 | 1.846 | 1.877 | 2.439 |
| 1993 | 13,281.5 | 2.116 | 1.604 | 1.857 | 2.405 |

SOURCES: Interview data; Wulanaodu Gacca official document 1984–1993.

entire growing season. At such low output levels, each adult cow is said nutritionally to require thirty-five mu of pasture, with each sheep and goat requiring seven mu (Wengniuteqi renmin zhengfu 1988). A look at Table 6.3, however, reveals that in Wulanaodu, summer-fall stocking ratios on unenclosed land have not fallen below 3.0 since decollectivization, climbing as high as 5.96 in 1989. This figure is more than 2.7 times the estimated critical stocking rate.

Second, the data demonstrate how it is possible for seasonal stocking ratios (and thus grazing pressure) to increase significantly over time without any remarkable change in livestock numbers. In recent years, at least, relentless grazing pressure derives not so much from ever larger numbers of animals but from the effects of a shrinking land base during the summer and fall. As the total land area of private household enclosures increases yearly, grazing pressures from late spring to late fall intensify on the unenclosed public range, even though total SEUs in the village may remain steady, increase somewhat, or even decline. For example, since 1990, the stocking ratio on unenclosed land during the summer and fall (Table 6.1) surpasses that of 1984 despite the fact that sheep equivalent units are generally down by an average of 5 percent.

Third, late spring and early summer grazing seasons on unenclosed range can no longer be assumed to provide ample nutrition for livestock trying to recover from bitter winter months. The traditional animal husbandry production bottleneck of winter and spring feed has now, in a sense, been stretched into the summer months as well. Livestock in Wulanaodu are increasingly likely to experience nutritional distress for longer periods. This phenomenon manifests itself in recent years through local government decisions to extend by up to ten days the usual June 1 cutoff for open grazing on collectively enclosed hayfields.

Fourth, the data testify to what the majority of grassland residents already know from experience. That is, in Wulanaodu, any pre-reform imbalance between livestock forage demand and available forage has only been exacerbated in the post-reform years by an unregulated enclosure policy that drops all animals on a shrinking land base at the same time for the entire growing season year after year. Every last household manager I interviewed asserted that the productive capacity of unenclosed rangeland has declined significantly since privatization. At least 39 percent of those same people dare to assert openly—despite the legacy of political fear—that enclosures, as they function in their community, are not a good policy initiative from the collective point of view. This accords with the experiences of herders all across IMAR grasslands as reported by William Hinton: "No matter where we went the people all said the grazing was worse than it had been ten years ago and much worse than in their childhood" (1992: 107). If ground cover is crucial to impede effects of wind and soil erosion so integral to land degradation processes, then increased stocking ratios on large tracts of sparsely vegetated scrubland amounts to a grazing regime that deliberately puts the majority of land at risk in order to protect an occasional oasis of enclosed greenery.

Enclosures as implemented in Wulanaodu doubly jeopardize the wider ecosystem by training intense grazing pressure on the same vulnerable land throughout the entire growing season. Any potential relief that enclosures might provide the system through rotational grazing with public access land is largely diluted because of uncoordinated herding routines among households. Nor does rotational grazing occur within enclosed land on any large scale. When households manage to acquire fence-wire, they typically seek to expand their net holdings, rather than parcel up what they have by subfencing. Under traditional forms of mobile pastoralism, livestock overgrazing on any one pasture was done temporarily so that vegetation could regenerate. Under the imposed grazing system, the unenclosed range remains exposed to intensive grazing pressures all the time.

There are serious ecological implications from this uncoordinated management system. Prolonged intensive grazing can provoke dramatic effects upon a wide variety of arid-land ecosystem properties that have been well documented. With regard to grassland vegetation, Risser (1981: 332–379) reported changes in stature, transpiration, patch size, nitrogen uptake, and a host of other modifications, depending upon plant species. And contrary to what seems to be a common perception among grassland officials in China, intensive grazing during the growing season is just as harmful to the range as it is during spring and fall.

Soil studies conducted in eastern Inner Mongolia stress the need to allow grass to regrow at least fifteen centimeters between periodic grazing, as well as to prohibit grazing during the first twelve to eighteen days of spring growth, and for thirty days at season's end, to maximize vegetation performance and protection (Liu Yuchen 1990). Soil carbon levels may decline by as much as 15 percent when 50 percent of plant production is removed during summer grazing, with losses estimated to be even higher on sandy Chifeng soils (Chuluun et al. 1993). Furthermore, aboveground biomass is believed to decrease "heavily" when grazing intensity during growing periods approaches 40 percent of plant growth, with additional detrimental consequences for soil salinity, pH, nutrient cycling, and organic matter (Kou et al. 1993). Loss of soil cover from intensive grazing is particularly detrimental in eastern Wengniute where wind speeds are so constant and strong. Sheet and gully erosion from water constitutes a significant hazard as well, despite arid precipitation levels. Rainfall tends to be concentrated within two or three months, causing frequent summer floods. Given these considerations, the fact that current grazing practices overload the open range throughout the entire growing season clearly violates the conservationist principles of rangeland science. Yet officials and scientists in China do not seem to appreciate the fact that if large tracts of enclosed land are fallowed every year for the entire growing season, then remaining unfenced areas must bear the burden.

The ratios listed in Tables 6.2 and 6.3 should not be construed as absolute values, given the unavoidable potential for informant error in the estimation of either livestock or enclosure size. Nevertheless, they do constitute the best available record, and accurately reflect upon a phenomenon that hundreds of residents throughout the township confirm to be true. The figures are certainly reliable enough to demonstrate the essential trend of a grazing management strategy that is contributing significantly to ecosystem decline over a time frame of sufficient duration to merit serious concern.

### Policy Adjustments

In tacit recognition of some of these problems, an adjustment in official grassland management policy emerged in 1988. By then, some Chi-

nese grassland specialists began to acknowledge that grassland reconstruction could not keep pace with rangeland degradation (see Ba 1993:
19). Subsequently, officials have enthusiastically promoted a modified
policy, dubbed the "small enclosure policy" (*xiaocao kulun*), which encourages independent households to devote their resources to more modest and practical goals than desert fixation, like intensive production of
hay and fodder crops on enclosed plots of land no bigger than twenty to
thirty mu. The "small enclosure" would serve as an artificial pasture,
bringing together for maximum productivity five key elements: water,
grass, grain, trees, and equipment.

The plan calls first for sinking private wells and constructing a system
of water conservancy. Next, leguminous grasses should be planted on 40
percent of the land area, with another 30 percent dedicated to silage grain
(usually corn), and 20 percent reserved for other grain feed (usually soybeans). A shelter belt of poplar, willow, or elm trees should occupy the remaining 10 percent of land area, both as a means to manage enclosure
microclimate and to diversify family economy with tree farming. Finally,
electric pumps operating four to six hours a day should complete the set,
but additional recommended equipment includes sprinklers, tractors,
harrows, hay balers, and other processing tools.

The program is believed to be a quick and efficient way to raise technological levels and managerial skills where they are sorely lacking
among "backward" communities. Implementation of these goals, however, remains subject to the initiative and financial capabilities of each
household. According to optimistic officials, after an initial downpay-
ment of 2,500–3,000 yuan (roughly U.S.$300–375) toward the purchase
of fence-wire, seeds, tree saplings, and water pumps, 60 percent of the investment is recouped within the first year of production, and profits ac-
crue by the second year.

Whether actual productivity can meet these high expectations or not,
the fact remains that not many families can afford such expensive startup
costs. And despite policy plans to make available benevolent loan options, credit is not extended to many people in rural China. As a result of
poor financing, the artificial pastures rarely work according to plan, often missing two or three elements from the intended set of five. Even the
crucial component of grass production seldom meets specifications, as
households typically forgo managed alfalfa cultivation and make do with
what grows naturally. In Wulanaodu, the best managed artificial pasture
yields about 370 kilograms of haystraw per mu, an admirable harvest by

local standards but far short of the projected 541.9 kg/mu for heavily fertilized fields (see Hu 1990: 216).

## Desperation Fences

The Chinese have an expression that translates as "going out empty-handed to capture wolves" (*kong shou tao bai lang*). Rural people say this usually in the context of inadequate or improper preparation to tackle a difficult task. In Nasihan, some residents use it to mock grassland policies that call upon common herders to fix moving sand dunes and check advanced wind and soil erosion processes without the benefit of investment capital or even meaningful institutional support. For the majority of anxious herders who cannot exercise their land use rights, the smug enjoinder to flourish through the means of small enclosures amounts to a government invitation to expose themselves to the twin wolves of blowing sand and plundering neighbors.

In the race to enclose fertile rangeland, the most limiting factor of production for many households has become wire fencing. Without it, there is no way to assert one's rightful place on the landscape, no way to prevent neighbors from grazing (or even enclosing) the land one has been allocated. Yet this valuable commodity is not so easy to obtain. In 1994, high-quality barbed wire cost about seven yuan per kilogram. Even those households who could afford such an expense had difficulty locating it in large quantities, even in large metropolitan areas such as Chifeng City. Often, elite herders resort to the purchase of raw wire (at 3.6 yuan per kilo) and then hire local people to weave it into sturdy netting. A cheaper fence material made from recycled wire mesh has recently become a popular alternative and can be purchased in local stores (when available) at 3 yuan per kilo. The material, however, has the feel of plastic and provides only a weak barrier, so it tends to mark the landholdings of less financially secure households. No private households set up durable concrete fence posts. Instead, residents simply impale into the ground a collection of tree branches and wooden planks of various heights and widths and structure. The cumulative effect of such haphazard fence design is to create a public montage of wood and wire that conveys the personality of the place—a certain stubborn simplicity that smacks of resistance to regimentation (see Figure 6.2).

Local fences come in many interesting varieties. Those households at the lower end of the economic spectrum have begun to pursue more desperate measures to maintain the integrity of their landholdings. Without

FIG. 6.2. Wulanaodu fences create a haphazard montage of wood and wire

proper fence-wire, they try to create makeshift obstacles to dissuade live-stock from roaming across their boundaries (see Figures 6.3–6.4). Some of them spend days at a time digging a deep trench in the ground, which quickly disappears with the forces of erosion. Others combine trenches with wire and sticks, which is also ineffective. Still others simply mark their place on the landscape with a single strand of wire—or even string—which does not keep animals or neighbors from passing over it but does mark the land as "occupied." Even more poignantly, some de-termined households have started claiming their share of land by simply thrusting broken sticks into the outlying dunes, hoping to reserve their stakes in the land before nothing of value is left.

In this context, it is interesting to note that two conflicting precedents of traditional (nineteenth-century) land use have been exposed and ag-gravated by the privatization and parcelization of the reform era. On the one hand, the traditional grazing system protected the principle of open range. As previously discussed, customary law gave common herders the rights to graze basically wherever they pleased within the confines of their banner. On the other hand, a customary practice did recognize some rights to occupy specific campsites exclusively, especially through the winter. The way to achieve the rights of "usucaption" was to impose

FIG. 6.3. Trenches create obstacles for livestock

FIG. 6.4. Makeshift fences built of scavenged materials

some built structure on the landscape, such as a dwelling, a well, a burial ground, or an animal pen. Customary law prohibited herders from using structures they had not helped to build (Szynkiewicz 1982: 23). Even today, Nasihan residents will not draw water from a well they did not help to construct.

Along with population growth, grassland enclosure policies have helped to pull these somewhat contradictory historical precedents to a point of maximum tension. The traditional rights of access to pasture and free mobility that are still cherished by all residents have been checked by the secondary rights of usucaption, now fortified by law under the ideologies of decollectivization and "grassland reconstruction." Barbed-wire fencing constitutes the necessary stamp of built structure that historically entitles the first-comer to special use privileges. It fits the culturally sanctioned principle of winter camp, though enclosed areas now often exceed the limits of a reasonable claim. In light of this cultural nuance, it becomes easier to understand how elite residents of Wulanaodu have managed to bully their neighbors with expanding enclosures with so little organized resistance. It helps explain why poor residents mark their claims on the land with flimsy materials until they can afford fence-wire. It helps explain why de facto occupation of the range is usually a more relevant consideration in settling land disputes than the wording of formal household proprietary contracts. More broadly, it helps explain how the intrusion of a new technology like fencing, which is motivated from outside township borders, could be so manipulated by local social realities to acquire such powerful local meaning.

# SEVEN

# Enclosure and Changes in Social Landscape

Access to sufficient quantities of grass and fodder constitutes the essential foundation for animal husbandry. By initiating a process that has led to the inequitable redistribution of wealth in landed assets, policies of enclosure in Inner Mongolia have simultaneously dictated the inequitable redistribution of other assets, such as livestock and capital. The poorest residents who cannot protect their land clearly bear the brunt of ecosystem decline. A downward economic spiral slowly envelops them, and they find themselves increasingly powerless to prevent their share of community resources from declining with each passing year. It is therefore not very surprising that since decollectivization, transformations in the social landscape have emerged or intensified that are every bit as dramatic as those of the physical landscape.

## Economic Stratification

Since decollectivization, economic stratification has been growing in China at a national, regional, and local level, and a nascent class structure and ethos have superseded the former egalitarian ideology and practices that once characterized the collective era. Knight and Song (1993: 197) report that interprovince income inequality accounts for 39 percent of total stratification among all Chinese counties, while intraprovince income inequality accounts for 61 percent of the differences. A Chinese government newspaper has reported that in rural areas, the average wealth share ratio between the top 20 percent and bottom 20 percent of households grew from almost a factor of three in 1981 to a factor of five by the summer of 1994 (*China Daily* 1994). On average, therefore, the wealthiest one-fifth of rural households now control roughly five times the wealth controlled by the bottom one-fifth. The data indicate, however, that residents of Nasihan have experienced even more extreme wealth stratification than the national average over the same period.

During 1993–94, I gathered detailed economic data for each household in Wulanaodu village since decollectivization, and one of the most important indicators of relative household wealth was livestock holdings. Of course, livestock holdings are not a complete measure of either wealth or income, because they do not necessarily indicate the transferability of wealth between livestock and other capitalized assets, nor do they necessarily measure the earned wages that have recently become a more salient economic factor in the village (especially with the increase of migrant labor over the last few years). Nonetheless, they are important, since 87 percent of village income still derives from animal husbandry. Also, residents told me they are not comfortable maintaining large sums of cash, and many do not trust banks, so they still prefer to convert economic value into livestock, accumulating animals as money "on the hoof."

I also analyzed two other indicators of household wealth for the entire village: tax liability and fixed capital assets. Taxable assets include livestock, landholdings, and household labor, so tax liability provides a general measure of economic well-being. Fixed capital assets include heavy equipment, fence-wire, and housing, and the village accountant keeps a running assessment for each household of such critical assets. I was granted access to the official government figures for livestock, household tax liabilities, and fixed capital assets for 1993.

Tables 7.1–7.3 summarize the data for each of these three indicators. The tables all group households according to two variables, production level (the number of livestock, amount of tax liability, or value of capital assets) and relative wealth bracket (such as top 20 percent or bottom 20 percent). Taken together, the tables show that, no matter what measure of wealth is used, households within the village emerge as clearly stratified. Table 7.1A in particular indicates that only 8 percent of village households control 200 or more sheep equivalent units, whereas 63 percent control fewer than 100 SEUs. It also indicates that increases in livestock holdings are positively correlated with average household enclosure size. The top 5 percent of households control average enclosure sizes of 60 hectares, whereas the bottom 25 percent control an average of less than 6 hectares.[1] As sheep equivalent units increase, the average annual enclosed landholdings for corresponding families also increases. Table 7.1B further indicates that by 1993, the top 20 percent of households account for 42 percent of community livestock holdings, whereas the bot-

TABLE 7.1A

*Household Distribution of Village Livestock, 1993*

(SEU)

| Sheep equivalent units | Number of house- holds | % of households | Avg. annual household enclosure size |
|---|---|---|---|
| 0–49 | 42 | 26% | 5.7[a] |
| 50–99 | 59 | 37 | 11.9 |
| 100–149 | 38 | 24 | 13.6 |
| 150–199 | 8 | 5 | 28.3 |
| 200–249 | 5 | 3 | 34.0 |
| 250+ | 8 | 5 | 60.1 |
| TOTAL REPORTING 160 | | 100% | |

SOURCE: Data for Tables 7.1–7.3 derived from interview data; Wulanaodu Gacca official document 1993.
[a]The enclosure size for households owning fewer than fifty SEUs is inflated somewhat by the values for two anomalous households included in the category. Both enjoyed large landholdings from as early as 1984, but their herds declined in the 1990s owing to family health problems. Discounting these outliers, we arrive at a figure of only 1.8 hectares.

TABLE 7.1B

*Household Distribution of Village Livestock, 1993,*
*by Wealth Strata*

| Percentile rank within community | Range of SEU | % of total herd |
|---|---|---|
| 1–20 | 128–300+ | 42% |
| 21–40 | 90–127 | 23 |
| 41–60 | 89–66 | 17 |
| 61–80 | 66–42 | 12 |
| 81–100 | 41–0 | 6 |

tom 20 percent account for only 6 percent of the total herd, which translates into a 7-to-1 wealth share ratio.

A similar pattern emerges for the full range of taxable assets, including land base, livestock, and family labor power. As Table 7.2A indicates, only 9 percent of village households bear an annual tax liability of more than 600 yuan, whereas 24 percent are liable for less than 200 yuan. Table 7.2B further shows that the top 20 percent of households account for 41 percent of the total tax base, whereas the bottom 20 percent account for only 6 percent, which again translates into a wealth share ratio of roughly 7 to 1.

These figures align with stratification of household wealth in other capitalized assets, including machinery, fence-wire, and housing. Only 8

TABLE 7.2A

*Household Distribution of Taxable Assets, 1993,*
*by Tax Assessment*

| Tax assessed (yuan) | Number of paying households | % of paying households |
|---|---|---|
| 0–99 | 15 | 9% |
| 100–199 | 26 | 15 |
| 200–299 | 46 | 27 |
| 300–399 | 35 | 20 |
| 400–499 | 17 | 10 |
| 500–599 | 16 | 10 |
| 600–699 | 6 | 3 |
| 700–799 | 4 | 2 |
| 800–1,100 | 7 | 4 |
| TOTAL REPORTING | 172[a] | 100 |

SOURCE: Data for Tables 7.1–7.3 derived from interview data; Wu-lanaodu Gacca official document 1993.

[a]In 1993, the village officially registered 174 resident families. The discrepancy in numbers of reporting households between various tables derives from two principal reasons: (1) Some of the families are transferred government cadres, who own taxable assets but are not allowed to keep livestock, since they are not entitled to landholdings. (2) A few households are recorded inconsistently across different categories, usually because they consist only of an elderly individual who actually dwells within another household.

TABLE 7.2B

*Household Distribution of Taxable Assets, 1993,*
*by Wealth Strata*

| Percentile rank within community | Range of tax liability | % of village tax base |
|---|---|---|
| 1–20 | 1,100–490 | 41% |
| 21–40 | 489–358 | 25 |
| 41–60 | 357–267 | 18 |
| 61–80 | 267–178 | 10 |
| 81–100 | 177–0 | 6 |

percent of households enjoy assets valued at more than 10,000 yuan, whereas 35 percent hold assets valued at less than 3000 yuan (Table 7.3A). Slicing the same data differently, we can see that the top 20 percent control 44 percent of village capitalized assets, while the bottom 20 percent control only 5 percent (Table 7.3B). Measured in terms of total capital assets, the wealth share ratio would be 8.8 to 1.

Of course, an important question remains: How much economic stratification already existed in the village at the time of decollectivization? As best as I can piece the situation together with available data and inform-

TABLE 7.3A

*Household Distribution of Capitalized Assets, 1993,*
*by Total Assets*

| Capitalized assets (yuan) | Number of households | % of households |
|---|---|---|
| 0–1,499 | 18 | 11% |
| 1,500–2,999 | 39 | 24 |
| 3,000–4,499 | 48 | 29 |
| 5,000–6,999 | 22 | 14 |
| 7,000–9,999 | 22 | 14 |
| 10,000–13,999 | 5 | 3 |
| 14,000–22,000 | 8 | 5 |
| TOTAL REPORTING | 162 | 100% |

SOURCE: Data for Tables 7.1–7.3 derived from interview data; Wu-lanaodu Gacca official document 1993.

TABLE 7.3B

*Household Distribution of Capitalized Assets, 1993,*
*by Wealth Strata*

| Percentile rank within community | Range of capitalized assets | % of village assets |
|---|---|---|
| 1–20 | 2,000–7,000 | 44% |
| 21–40 | 7,000–4,000 | 22 |
| 41–60 | 4,000–3,000 | 17 |
| 61–80 | 3,000–2,000 | 12 |
| 81–100 | 2,000–0 | 5 |

ant recollection, the community was economically stratified in 1981 at a level consistent with the national rural mean, with a wealth share ratio of 3 to 1 or lower. Both land and livestock (the crucial factors of production) were originally distributed among village households with reasonable equity in the early 1980s. Since virtually no two villages in IMAR implemented decollectivization in exactly the same way, it is necessary to clarify what actually happened in Wulanaodu before assessing the available data.

Land use rights to rangeland resources (*muchang*) were distributed on a household basis. Each household obtained a contract providing for roughly 700 mu of land (47 hectares),[2] though the actual number varied with specific cases. A few households received portions as large as 1,500 mu while some acquired as little as 500, depending upon the overall quality of rangeland. I never met any resident who grumbled about the principle of allotting differential sizes of landholdings in the redistribution process because of land quality. Some people did grumble, however, about personally receiving inferior locations on the landscape. Judging land as "inferior" involved the consideration of many variables, including the quality of vegetation, centrality, and *feng shui* attributes.[3] Land use rights to hayfield resources were also distributed evenly, based upon livestock holdings and household labor power.

The distribution of livestock was calculated on a per capita basis and also attempted to achieve rough parity. Many residents, however, still grumble that certain households did gain unfair advantages in terms of quality of livestock, or in terms of acquiring an extra steer. No doubt, exact parity was impossible to achieve, even without the problems of favoritism. But the process worked reasonably well. I managed to gain access to village records that revealed how many animals were actually distributed to each independent household in 1981, and they indicate a fairly equal starting point in terms of livestock. The data are summarized in Tables 7.4A–B.

As Table 7.4A indicates, no household controlled more than 150 sheep equivalent units, and 32 percent controlled either fewer than 50 or slightly more than 100 SEUs. The bulk of the population (two-thirds of all households) controlled between 50 and 100 SEUs. The breakdown by relative wealth bracket in Table 7.4B shows a wealth share ratio of only 2.6 to 1. Thus, Wulanaodu households started the reform era with relatively modest stratification of contracted land and livestock holdings. Decidedly unequal, however, were the crucial variables of social privilege that determined household access to other collective assets like fence-wire, equipment, bank loans, and political sanction.

Tables 7.1A–7.4B help to quantify what every resident in Nasihan already knows: that household shares of village wealth have become increasingly stratified since decollectivization. At least in part, this can be attributed to the proliferation of fence-wire, which instituted de facto unequal distribution of access to essential community resources. Of course, animal husbandry is inherently a risky enterprise, especially in such a marginal environment where both chance and skill play a role in herd performance. Some increased stratification would be expected after more than a decade of privatization. Also, other differences in labor power, motivation, and skill would obviously contribute to increasing stratification. Nonetheless, considering the many exploitative practices associated with rangeland parcelization in Nasihan, enclosures have undoubtedly accelerated the pace and magnitude of more natural polarizing mecha-

TABLE 7.4A
*Household Distribution of Village Livestock, 1981,*
*by SEU*

| Sheep equivalent units | Number of households | % of households |
|---|---|---|
| 0–49 | 24 | 23% |
| 50–99 | 71 | 68 |
| 100–149 | 10 | 9 |
| 150–199 | 0 | 0 |
| 200–249 | 0 | 0 |
| 250+ | 0 | 0 |
| TOTAL REPORTING | 105 | 100% |

SOURCE: Data for Tables 7.4A and 7.4B derived from Wulanaodu
Gacca official document 1981.
   NOTE: This table does not provide complete household livestock
holdings for 1981 because small private herds were maintained prior to
decollectivization, and a registry of total livestock holdings for all village households at the end of 1981 does not exist. Nonetheless, by interviewing reliable informants, I ascertained that these private herds
were quite small (usually involving no more than a few sheep, goats, or
cows), so that any prior disparities in livestock holdings were minimal.

TABLE 7.4B
*Household Distribution of Village Livestock, 1981,*
*by Wealth Strata*

| Percentile wealth rank in community | Sheep equivalent units | % of total herd |
|---|---|---|
| 1–20 | 126–90 | 29% |
| 21–40 | 89–73 | 24 |
| 41–60 | 72–63 | 19 |
| 61–80 | 62–49 | 17 |
| 81–100 | 48–0 | 11 |

nisms. Indeed, given the rather elastic and laggard relationships between
the variables of fence-wire, rangeland vegetation, livestock production,
and asset accumulation, it is striking to see the consistency with which a
variety of different data sets attest to new levels of stratification.

## Stratification of Living Standards

Changes in patterns of wealth distribution have quite obviously pro-
voked changes in the distribution of daily living standards among community residents. Housing and nutrition provide two immediate examples. My frequent home visits gave me the opportunity to observe the

range of housing and nutrition throughout the township, particularly because Mongol norms of hospitality require hosts to provide guests with food and drink. As a "foreign dignitary," I was generally admonished by each household to linger for a meal or refreshment. I can thus report that while some prosperous families have managed to build new and impressive brick homes in recent years, the majority of residents continue to dwell among the dirt floors and drafty walls of mud-brick (see Figures 7.1 and 7.2). While some prosperous households enjoy a relative abundance and variety of foods, many others make do with only two daily bowls of rice and a few dehydrated vegetables through winter months. According to several knowledgeable residents, at least 10 percent of village households are not getting enough to eat every day, especially during spring season when supplies have run low and livestock are too lean to sell for cash. The problem was particularly acute in April 1994, when the local price of rice jumped unexpectedly from 0.7 yuan per jin to 1.1 yuan per jin (1 jin = 1/2 kg).

Differential household income also redistributes access to government services in education and health care. To obtain schooling beyond the third grade, children must relocate to the township central village some 10 kilometers away, about a half-day's journey by mule litter for most Wulanaodu residents. The geographic distance requires that families pay room and boarding fees in addition to tuition costs (for a total of roughly 100 yuan per semester per child). This is an extravagance that many cannot pay for even one child for more than a few years. I interviewed many struggling parents who were embarrassed to inform me that they could not afford schooling for their children beyond the third grade. Beyond the sixth grade, children must relocate to the more expensive town of Hong Shan, so distant as to be inaccessible without cross-country bus service. In contrast, a few of the most prosperous families enjoy great prestige for sending their teenage children as far as Wudan (the county capital) for higher education. One high-ranking party member even managed to enroll his daughter at a national university in distant Hohhot. For him, the requirement that all children beyond the third grade attend a boarding school in the township capital served as a benefit to village residents, because it meant that children would have fewer distractions: "The more difficult their situation, the better for them because adversity instills discipline and makes them appreciate hard work."

Residents face the same geographic and financial obstacles to basic health care delivery. For rural China in general, economic reforms may

FIG. 7.1. New, expansive brick home

FIG. 7.2. Typical mud-brick home

have brought new prosperity and improved health, primarily by an increase in disposable income that offsets shrinking institutional supports (Knight and Song 1993). But in marginal areas, decollectivization has meant a decline in access to affordable and quality health care. This is because in the early 1980s, cooperative medical care systems disappeared in all but the wealthiest rural areas. Townships assumed control of commune facilities, and medical personnel began to charge on a fee-for-service basis geared toward maximizing profits. In Nasihan, the poorest residents increasingly rely upon their own home remedies for immunization and medical care.

Even if Nasihan residents do seek professional attention, a visit to the single township clinic typically requires losing a full day of labor just for transportation. Medicine can be purchased, but little can be done for serious illness or traumatic injuries. Those cases are forwarded to county hospitals some four to six hours away by irregular bus service. After combining the expenses of medication, transportation, and lost labor, many families cannot entertain the thought of health care, let alone health maintenance. Increasingly, access even to the most basic government services has become a matter of economic privilege.

I can illustrate the gravity of this situation by recounting the story of a young male herder, Renchin, who has been physically disabled since the age of fifteen. To this day, nobody likes to talk with him because of a lingering stench that villagers attribute to the stump of his left leg. In the winter of 1982, Renchin and a friend prepared to make a hay delivery to distant pastures where family flocks were stationed to browse the roots of a dormant range. Before setting out, they decided to fight the cold by sharing a bottle of liquor. Although drunk, they set out and soon mismanaged the load they were transporting so that cow, passengers, and hay all toppled violently to the ground.

The cow was pinned on his horns from the weight of the cart and slowly froze to death that night. The two passengers were not able to right the cart and could not rouse themselves from the cold earth before falling asleep. The friend was protected on all sides by hay, but Renchin fell in a position that left a hand and leg exposed to the winter winds. When the friend awoke sometime later, he discovered the severity of Renchin's loss of body heat and somehow found the strength to carry him to safety. Over the next several days, Renchin suffered pain from frostbite in his left hand and foot, and the odor of rotting flesh became a topic of conversation throughout the village. Though the brother of a township-level cadre, he was without medical insurance and too poor to pay the transportation and medical fees involved in a journey to the nearest hospital. Slowly his limbs turned gangrenous, and left to his own devices, he saw no alternative but to amputate his own fingers and foot with an ordinary butcher knife. Villagers still talk about the howls that rang through their alleys that afternoon.

## Labor Commodification

The increasing concentration of wealth has many direct implications for other changes in social organization and daily practices within herding communities. One important emerging development is the increase in outmigration of labor. Since 1990, more and more downwardly mobile households have started sending family members into distant counties to earn wages at menial labor, usually in unhealthy settings such as brick factories. By 1993, township officials estimated that 5–10 percent of the population was venturing away from their familiar ethnic surroundings, contracting themselves out as unskilled labor for six months at a time. In Wulanaodu village, thirteen different households sent out family members as migrant laborers in 1993. Each of these households belong within the two poorest categories shown in Table 7.1A (that is, their SEUs fall below 100); all belong within the four poorest categories of Table 7.2A (their taxable assets fall below 400 yuan); and all belong within the three poorest categories of Table 7.3A (with capitalized assets below 5,000 yuan).

The outflow of manpower helps prop up the declining pastoral economy, both by increasing the revenue of needed cash and by reducing the population pressure dependent on local resources. The experiences, however, are rarely enjoyable or even positive. A typical example of the exploitation that may occur is recounted in the following abbreviated narrative, as told to me by one bitter resident.

In 1993, Deligen heard that a brick factory in Wudan was hiring, so he left his home in the company of two others to seek extra income. He agreed to push a brick cart at the rate of 0.18 yuan per load for two months, to be paid upon termination. After one month, the wage rate was suddenly cut to 0.14 yuan per load. The boss told Deligen that he was free to leave, but "no back wages would be paid to quitters." So Deligen worked three months without terminating, then decided to ask a month's leave at the end of August to help cut hay in Wulanaodu. The boss excused him, but granted him only a week's leave. Yet Deligen stayed away for the entire harvest month. Upon his return, the boss said nothing about it and immediately put him back to work. After two more weeks, Deligen decided to terminate and so he asked for all his back wages. Only then did the boss inform him that the brick factory had fined him for the slow pace of production during the three weeks of his absence. His "fine" exceeded the back wages due him, so the boss refused to pay anything at all. Not knowing anyone in the city, and not having the language skills to press legal charges, Deligen returned home empty-handed after more than three months of hard labor.

Labor patterns among village residents who remain at home have also changed. Increasingly, poorer households sell their labor to wealthier households. Environmentally privileged households find themselves more and more in need of labor to keep up with their expanding production. They hire neighbors to help them herd animals, cut hay, harvest fodder crops, shear sheep, comb cashmere, construct homes, and even to weave wire into more fencing to expand their enclosed land base still further. Some households have even begun hiring neighbors to cook and clean house. In contrast, land-poor households face unwelcome cutbacks in their own production, and so find themselves forced to work for wages to make up for insufficient livestock. Their labor power is thereby co-opted to promote the very forces of social power that unjustly consume their own productive assets of land and livestock. Particularly troubling are the stories of skilled adult herders who once cared on their own lands for the animals of village elites, only to lose their clients as soon as their unprotected pastures became too degraded to maintain satisfactory levels of productivity. In this way, wealthier households literally eat the poor out of house and home, then move on to the next victim.

There have been other changes in local labor patterns. Given the increased social and ecological risks of production faced by independent herding households since decollectivization, one might expect the *lianhu* system to thrive. Instead, enclosure competition has soured community relations so that households have increasingly atomized their production operations and commoditized their labor exchanges. Many formerly successful *lianhu* dissolve because allied households simply lack confidence in their herders to deal honestly with them in managing their livestock and other assets. Most interhousehold labor cooperation is now conducted between close family relations, such as brother to brother or father to son. When informal labor swaps do occur outside that context, usually during September and May, they are quite limited in duration (a day or two) and quite specific about reciprocal obligations between parties.

## Production Diversification

Closely related to increasing stratification and expanded commercial opportunities for local residents, there have been significant alterations in local practices of resource management. While Wulanaodu herders do share the dream of becoming rich, diversification is primarily motivated by more conservative considerations. As land degradation intensifies on the range and the ideology of collective security vanishes, household managers perceive the need to spread increasing economic and environmental risks among a variety of resources and production strategies. Diversification is a common strategy of coping with unpredictable variability in pastoral production when reliable access to a market exists (see Barth 1966; Jochim 1981: 91–92).

The attempt by modernizing national economies to harness more fully the productivity of pastoral peoples generally permits two broad options at the local level: residents can specialize to expand their commercial production, or they can diversify to supplement their core economy. In Wulanaodu village, there is some evidence that both strategies are in operation. Many households are attempting to streamline their animal husbandry production by marketing more animals. A few households even decided to sell most or all of their livestock in order to specialize in new economic activities such as transport services. One young householder bought a three-wheeled motorcart by which to conduct petty trade in bricks, coal, cigarettes, and liquor. By 1994, at least six different small supply stores had emerged just within the residential center, four of them operating out of their homes, and all competing to sell the same few items for a slender profit margin. Many residents practice petty trading in other ways without ever leaving the community. They may buy up grain supplies from local stores in order to sell it later to neighbors when supplies have run out. Some earn a few cents by purchasing fresh vegetables in town or at the research station and then delivering them to remote pasture homesteads.

Despite some attempts at economic specialization, the predominant strategy by far is for herding households to seek production diversification. Many household sideline operations have expanded since decollectivization, most vigorously since the 1990s. Since 1987, an important sideline has been intensive agriculture. Back in the late 1950s, the collective experimented with crop farming on a sizable scale. A group of herders was sent into nearby Han settlements to learn how to cultivate.

The banner government responded by transferring eight Han families into Nasihan to help them grow soybeans, corn, maize, millet, and oats. Over time, all but one of those families retreated from the community, and farming operations were gradually scaled back until they virtually disappeared. When a retiring cadre returned home to Wulanaodu in the mid-1980s, he was dismayed to see all the fields reverted to hay pasture, so he organized about twenty-five "demonstration" households to sow grain. Since then, sideline farming has enjoyed an enthusiastic revival. According to existing regulations, cultivation is prohibited in areas prone to wind and soil erosion, or where annual rainfall is less than 335 mm. In Nasihan, residents are not supposed to sow more than 2 mu anywhere without explicit authorization. Despite these prohibitions, residents are now expanding their gardens into ever larger fields that often swell to six or seven hectares. They usually plant corn or soybeans, which may be sold or utilized at home as food or animal fodder. But in some areas of Nasihan, residents are starting to grow more labor-intensive and lucrative grains such as millet and rice. A few households have taken an interest in producing grapes as well, though no one harvests them on a scale large enough to profit.

Another growing sideline is fish farming, which county government officials have recently pushed as a development scheme. After receiving an initial loan of fish stock (usually about 2,000 fish), households farm and market the produce to repay the loan and earn additional cash. Many of the poorer households now stock desert ponds with fish, but lacking the tools to catch or market them efficiently, they rely upon urban traders who reap the bulk of the profit. Still, some residents claim to earn 500 yuan per year from this newfound production option. Likewise, tree farming has become a means to earn additional income. Usually, poplar twigs are planted in nurseries by the hundreds and sold to neighbors or to urban buyers for a few yuan apiece when they are harvested two or three years later. Other residents buy them to serve as windbreak and to sell as timber after maturing another five or ten years.

Consistent with a diversifying community, there have also emerged significant modifications within the practice of animal husbandry. One change is the increasing tendency for households to streamline their livestock operations. Most notably, fewer households keep large flocks of both sheep and goats. Traditionally, Mongols relied primarily on sheep and cattle, but today more herders prefer to specialize in goats for the following reasons: cashmere continues to be highly profitable relative to wool; goats are less vulnerable to cold, famine, and drought; goats consume a wider variety of forage; they require less labor to manage; and they impose fewer veterinary expenses. In Wulanaodu, the majority of herders now keep only a small flock of sheep for domestic consumption because they still prefer the taste of sheep to goat. Of course there are some households who have always specialized in sheep production and prefer to stick with it. They are happy to buy up local sheep to increase their own flocks as a hedge against future market fluctuations. But far and away, household flock structures favor goats, despite grassland regulations intended to limit their numbers to a modest ratio of three per family member. It seems reasonable to view this trend as an adaptive response to an environment increasingly dominated by sand and woody shrubs.

Another notable change is the general decline of animal husbandry expertise in the region. During the collective era, key households were selected to manage segregated livestock for the entire collective. Roughly forty herding households lived out on the rangeland and specialized in the care of a single animal type. Different animals have different grazing habits, so efficient management generally requires species segregation. Though the herders never received any special training to go along with their responsibilities to the collective, they did eventually acquire specialized knowledge and skills. With rangeland parcelization and a decline in the *lianhu* system of labor cooperation, however, routine chores of herd management have increasingly become the responsibility of each independent household. As a consequence, herding expertise has been diluted, mixed species typically graze together, and some critical rangeland management techniques (such as rotational grazing) have regressed to the lowest common denominator.

## Community Fragmentation

In association with the emergence of social stratification, increasing incidents of crime and violence testify to declining community solidarity. Enclosure disputes have motivated many neighborly feuds, even between families or friends who once enjoyed close and cooperative relationships. I can indicate some measure of this phenomenon by describing the wide variety of circumstances in which serious boundary battles have been known to occur.

First, heated arguments arise practically whenever a new fence goes up, if for no other reason than that property boundaries were previously never clearly identified. Private contracts issued at the time of decollectivization refer only to boundary outlines roughly demarcated by indistinct landmarks such as sand dunes, hills, or treestands. Even the boundaries separating different villages have no clear line of reference. Wulanaodu households residing along all four border areas of the village have required boundary dispute arbitration from township and even banner-level authorities.

Second, vandalism or outright theft of fence-wire frequently occurs, especially just after a new enclosure is erected. The research station lost more than a hundred meters of wire in a single theft during autumn 1993. Likewise, several hundred meters are repeatedly stolen from enclosures bounding the collective hayfield. Vandalism frequently occurs on privately controlled property, as well. Neighbors generally interpret enclosure expansion as a form of competition, responding venomously to vent their envy or insecurity. Thieves may use stolen wire for their own homestead, but they more often sell it to avoid detection. Contrary to some reports (for example, Longworth and Williamson 1993: 313-321), I found that fence-wire is a highly liquid asset on the Wengniute grasslands, easily reconverted into either cash or livestock.

Nor is theft limited to fence-wire. I also heard about significant burglaries of livestock, cashmere, vegetable produce, cash savings, timber, and even animal dung. The local perception of escalating theft, in combination with poor legal protection, contributes significantly to conservative resource management strategies. For example, herders are reluctant to assume responsibility for large sums of cash or equipment, so they generally avoid entrepreneurial initiatives that might enhance their terms of trade at the market or diversify their sources of income. With just a little organization and community trust, local herders could pool their capital long enough to benefit collectively from better terms of trade available in the city. Unnecessary dependency upon urban traders results in continued extraction of surplus value away from the community. Many local informants point to animosity generated by enclosures and rising theft to explain why they cannot organize local labor or other resources for mutual benefit.

Third, heated arguments often arise over the attempt of one neighbor to use another's fence line for his own enclosure. Sometimes, the dispute can only be settled by running adjacent parallel fences across the same field. Conditions on the range are so volatile that some neighbors prefer this wasteful outcome to joint ownership of wire, as Figure 7.3 shows.

FIG. 7.3. Adjacent parallel fences constructed to settle boundary conflict

Fourth, many conflicts begin when one neighbor actually circumscribes the gate of another, thereby cutting off all routes of entry to an interior private pasture. Similarly, many fences have been vandalized by disgruntled residents who are impatient to find a gate for passage. People simply do not honor the integrity of a fence if it offers no permeability for their transit. With some amusement I once found myself caught up in the process. After passing several hours conducting home visits, I returned to my motorcycle, which I had temporarily parked in a small field, only to find it newly surrounded by fence-wire that provided no access gate to the road. The only way I could exit was by uprooting a post to slacken the wire. I did not know for sure whether to interpret this action as intentional harassment or simply as the work of someone who had grown impatient with my parking. In any case, I felt no compulsion to restore the fence upon leaving and in that moment, I experienced a small measure of the frustration that many residents dealt with every day. Rangeland enclosure can be used to convey personal messages that sometimes provoke hostile feelings and behavior.

Fifth, animal incursions across enclosed territory especially tend to incite fury (see Figure 7.4). I learned of several incidents in which people brutally attacked a neighbor's trespassing animals, sometimes cutting off

FIG. 7.4. Animal incursion across enclosed territory

legs and horns. Animals injured in this way are usually slaughtered immediately, so the financial implications of such violence can be quite serious, especially when the damage occurs out of peak marketing season.

Finally, at least one additional form of enclosure conflict has just begun to surface, but it presumably looms as a larger problem in the not-too-distant future. Disputes will surely arise as households divide and more people move into pasture areas to claim land already enclosed and occupied by squatters. There is currently no clear precedent for resolving these predictable land tenure disputes.

Enclosure conflicts of one form or another have affected nearly every household in the village. According to numerous informants, the blatant display of neighborly discord over production issues that has emerged recently was rare, if not completely unprecedented, during the collective era. Of course, intermittent political campaigns generated a great deal of violence from time to time, but such chaos was generally instigated by outsiders and never became routine behavior. Other Mongol ethnographers have indicated that prior to the collective era, independent herding households managed to share the range "virtually without friction," despite the great potential for conflict generated by migratory grazing habits (Szynkiewicz 1982: 23). The desire to be respected and the need for labor alliances during peak production seasons helped to preserve social norms most of the time. A strong tradition of nonviolence prevailed on the range, even when selfish or antisocial behavior occurred (Mearns 1993: 80; see also Rasidondug 1975).

I do not mean to romanticize the social harmony of earlier periods, but the fact remains that contemporary residents of Nasihan complain incessantly about the rise of crime and violence since the beginning of land privatization. In addition, many of them despair about a more general decline in everyday social civility. One aged informant, for example, described his resentment when a neighbor asked permission to assume control of his land "since you are going to die soon anyway." Even village elites acknowledge that land parcelization has led to new patterns of social interaction, new crises in neighborly relations, and new levels of community fragmentation.

## Redistribution of Health Risks

Due to a host of climatic, geographic, sociopolitical, and cultural factors, the ethnic Mongols who populate the northern grasslands are susceptible to health risks of chronic cold stress and accidental injury and death from hypothermia. In the post-reform era, general susceptibility to such hazards has been mitigated somewhat by improved standards of living that raise levels of income, nutrition, housing, education, and hygiene for much of the population. At the same time, the traditional health risks have been retrained upon select community residents with new intensity, primarily because of differential patterns of alcohol use and abuse associated with the exploitative enclosure policy and the recent acceleration of income inequalities. The proliferation of household enclosures has not only exacerbated ongoing processes of soil erosion in selective areas. It has also, in a remarkable parallel shift, redistributed cold environment health risks disproportionately among the poorest families, especially adult male herders (see Williams 1997).

An unusually high number of ecologically underprivileged residents have come to experience the marginality of their land and their social status in direct bodily form, through maiming, amputation, or death following chronic and acute incidents of overexposure in a bitter environment. Resource-poor herders assume the greatest risks of accident primarily because of social conventions that structure the intensity and the circumstances of their frequent alcoholic indulgences.

Although village-level statistics are not kept for such accidents, a year of participant observation revealed to me that the incidence of limb deficiency and cold-related deaths are unusually high in the area. One of my earliest impressions of the community was the frequency with which I encountered people missing limbs or digits, or walking with canes. During household interviews, I personally met at least eight people who presented some form of limb deficiency. A veteran township doctor once told me that at least two or three deadly drinking accidents occurred every winter among adult males just within the confines of Wulanaodu and its two neighboring villages. Also, a large number of aged informants independently attested to the magnitude of cold injury health risks for active young men. Adjusting for the age structure of the population, the doctor's estimate would put the annual crude death rate from acute cold stress roughly between 2.88 and 4.32 per thousand men over the age of fifteen. That range compares with a national rural crude death rate due to "natural and environmental accidents" of only 0.01 per thousand men (WHO 1994: D-357). Unfortunately, no official estimate for the rate of cold-related injury is available at either the local or the national level.[4]

One herder actually froze to death very near the research station during my year of fieldwork. It is instructive to recount some of the details of his accident:

Tamujabu was found frozen amid crusted sand dunes on a December morning in 1993, not more than a half-mile from his mud-brick home. According to his surviving brother, romantic misfortunes sparked the lethal chain of events that ended his life at the age of thirty-one. After waiting two years to consummate a wedding engagement, Tamujabu and his fiancée found it impossible to finalize their marriage because local government officials authorized to register them had taken indefinite leaves of absence. (Unable to draw a salary for three months because of an "overheated" national economy that produced tight money in the provincial capital, many officials had resorted to moonlighting for private-sector wages in distant counties.) Tired of waiting, the fiancée gave up on the pending marriage and returned to her natal home. Tamujabu had spent his inheritance to obtain this bride, and so watched despondently as his financial investment—and last chance for domestic happiness—walked out of his life.

Tamujabu retreated along with his matchmaker to his sister's home, where they commenced some very heavy drinking. Like most Mongol men, Tamujabu was renowned for and proud of his robust drinking capacity. He could drink an entire sixteen-ounce bottle of *baijiu* (distilled

100-proof corn or sorghum-based liquor) with each meal, for a total of nearly forty-eight ounces in a normal day. No one knows exactly how much he consumed that particular evening before he set out, in strong winds and subfreezing temperatures, for a 7.5-mile walk home across blowing sand. He trudged a full seven miles before lying down to rest from exhaustion, but he never woke up. His stiff body was discovered the next morning by a distant neighbor while tending a flock of goats.

As local residents see it, Tamujabu did not die as a result of his own stupidity or carelessness, but as a result of the complacency of township government. In the poignant words of one village elder, "He died of anxiety, like many of us." This informant witnessed a death and looked beyond the liquor to see prevalent despair born from deepening poverty and the systemic capriciousness of government services. Indeed, it is telling that Tamujabu's family has become fearful, rather than angry, about what happened. They are anxious that local cadres might take some kind of preemptive action to silence their grievance, since it reflects so poorly on standard operating procedures within the township.

I interviewed the injured of numerous households (or their surviving relatives) in the area before realizing that they all shared closely similar experiences. In general terms, they were all active Mongol herders living in a cold and remote environment characterized by a dispersed residential pattern, limited access to health facilities, and a lifestyle regulated by social conventions that make their frequent alcoholic indulgences especially dangerous. All the injuries involved a night of social and excessive alcohol consumption, a long journey home through strong winter winds, and prolonged exposure that resulted in injury, frostbite, or death. Moreover, all the injured belonged to the poorest households in their community. I know of no cold-related injuries or deaths among the wealthiest one-third of village households who command the largest landholdings.

Household enclosure, in combination with general economic stratification, has disrupted local drinking conventions with detrimental effect for the poorest residents in a number of ways. First, greater commercialization in the countryside has made distilled grain liquor more conveniently available than ever before. Trade in liquor and cigarettes constitutes a new and major household sideline throughout Nasihan. A local store manager informed me that he now has trouble keeping sufficient quantities of liquor in stock, selling an estimated 2,300 *jin* (half-kilo bottles) per month. Apparently, levels of grain alcohol consumption have risen dramatically on a national scale in China since the beginning of the reform era. Production of grain alcohol nearly doubled between 1980 and 1987 (*Beijing Review* 1988: 7). And strong grain alcohol has now completely displaced the less potent traditional liquors that Mongol herders primarily used for sociability until just a few decades ago (see Lattimore 1994: 254–256). When I asked local residents about this transition, they explained that it requires too much effort to get drunk using the traditional beverages.[5]

Second, downward mobility increases the likelihood of alcohol consumption for several reasons. Alcohol creates a false sensation of warmth, which helps to compensate for a cold house made colder by inferior building materials, the high cost of coal, and scarce supplies of fuel. It also curbs appetite, which helps to compensate for insufficient caloric intake resulting from a decline in purchasing power.[6] Of course, residents from every social class also drink for the camaraderie and pleasure it brings. But for the poor, alcohol is often used to anaesthetize the sting of declining social status. During my stay, various informants independently asserted that drinking helped them feel warm, dull hunger, and "forget everything."

Third, downward mobility conditions the terms under which social drinking is most likely to occur. Relative economic power tends to dictate which individuals scramble most desperately for drinking partners, who plays guest, and who must make a long distance journey home while intoxicated. Resource-poor herders assume the greatest drinking liabilities in each of these three respects. With the advent of privatization and cutthroat land use practices, they have reason more than ever to seek risk-sharing alliances. Those who are consistently inclined to drink the most tend to be those who are most desperate for expanding social networks. Yet they are rarely in a position to play host to other members of the community, and due to long-standing cultural norms, the guest must usually drink to the point of inebriation. They then stagger away for a long journey home across blowing sand dunes in bitter temperatures. Earlier norms of hospitality used to include an invitation to sleep over until intoxication abated, but some residents told me that the practice has diminished. I infer the cause to be related to the erosion of inter-household cooperation and neighborly good will thanks to economic stratification and prevalent enclosure disputes. Most households still receive and serve uninvited visitors, but they do not feel compelled to shelter them under ordinary circumstances. Furthermore, since decollectivization, village authorities have permitted a more dispersed settlement pattern, so that homeward journeys are now more likely to require greater travel time.

In the absence of either collective authority or clear patron–client relations, the old customs that used to help control and minimize the risks of hypothermia no longer function as before. The poorest herders, after drinking heavily in another home, are now more likely to be turned out to stagger home on their own, and accidents are more likely to occur. Gone are the days when Jagchid and Hyer (1979: 117) could write that "one never hears of anyone falling off a horse because of drunkenness." Although any particular herder may frequently drink to excess, the risks they assume while intoxicated are structured in ways that benefit some at the expense of others.

The bodily injuries that land-poor male herders sustain are physical manifestations of the health risks they have disproportionately endured since the beginning of rangeland privatization. A chain of social process directly connects ongoing land degradation and bodily degradation. Select representatives of the community are literally made to wear their collective sedentarization in a bodily narrative that speaks at once of personal and community deprivation.

## *Spatiotemporal Alterations*

The proliferation of household enclosures in association with land reform and expanding rural commercialization has also triggered transformations in local experiences of space and time. Relative to their traditional pastoral orientations, Nasihan residents increasingly organize and conduct their daily lives under stricter and more tightly regimented spatial and temporal horizons. Increments of time and space were never absent in pastoral Mongol society, but the traditional increments were more expansive than those developed within industrial or even agricultural societies.

The long process of sedentarization over the twentieth century has slowly eroded many of the earlier behavioral differences between Mongol and Han, but neither the sedentary mindset nor spatiotemporal conformity has yet been fully achieved. Even in regions like Wengniute where sedentarization is relatively advanced, evidence of distinctive orientations still exists. Newly erected wire increasingly impedes the free movement of herds and herders across the once seemingly boundless terrain. Rangeland parcelization has thereby introduced a more structured and confining daily regiment, not only in terms of space but also in terms of time.

Spatial enclosure is visibly evident right on the landscape, but the corresponding notion of temporal enclosure perhaps requires a few words of explanation. With regard to traditional Mongol parameters of time, Jagchid and Hyer (1979: 114–115) make the following instructive observations:

In the popular mind, a nomadic life-style is liberated from time and schedules and the tyranny of clock-watching. There is a good deal of truth in this view, however, Mongols are more time conscious than foreigners might expect and, while they are not clock watchers in the Western manner, their concept of temporality should be given at least passing comment. Everyday life is timed according to the passage of the sun: on the steppe, people constantly watch their shadow, and within the yurt, the position of the sunbeam is observed as it comes through the ceiling and moves across the floor. Thus, activities of the day—meals, the handling of animals, and chores—are regulated.

I readily concede that even under the most nomadic conditions, daily Mongol production has always followed its own kind of temporal (and spatial) regularity. The point is, however, that such "traditional" orientations of time and space are relatively unconstricting compared to urban and farming communities.[7]

Even in semi-acculturated settings like Nasihan, the pace of life can be painfully slow to visiting outsiders. Daily life is still not divided into hours and minutes, but into three large chunks: morning (*zaoshang*), afternoon (*xiawu*), and evening (*wanshang*). Watches and clocks are present in the village, but they possess no meaningful authority. Timepieces are essentially decorative, banished to the interior of the house where they prove relevant only to the few evening hours of TV programming. Calendars hang in nearly every room of every house, but no resident knows what day of the week it might happen to be without first consulting it. Indeed, the entire township makes no distinction between any single day of the week. There are no "weekends," market days, or other regular schedules to impinge on the uniform march of time. The elementary school does close for short periods of recess (roughly three days every two weeks), but class schedules run by dates of the month regardless of the day of week. Unless adult residents have children in school, they are unlikely even to know what month it is.

The passage of time is still predominantly observed exclusively in terms of the agricultural season, which is organized into increments of a fortnight. Each year is divided into twenty-four distinct solar terms that help to codify the cycle of agricultural activity. Each of the four seasons

TABLE 7.5
*The Twenty-Four Solar Terms*

| | |
|---|---|
| *Spring* | |
| Beginning of spring (*lichun*) | Feb 3–5 |
| Rain water (*yushui*) | Feb 18–20 |
| Waking of insects (*jingzhe*) | March 5–7 |
| Spring equinox (*chunfen*) | March 20–22 |
| Pure brightness (*qingming*) | April 4–6 |
| Grain rain (*guyu*) | April 19–21 |
| *Summer* | |
| Beginning of summer (*lixia*) | May 5–7 |
| Grain full (*xiaoman*) | May 20–22 |
| Grain in ear (*mangzhong*) | June 5–7 |
| Summer solstice (*xiazhi*) | June 21–22 |
| Slight heat (*xiaoshu*) | July 6–8 |
| Great heat (*dashu*) | July 22–24 |
| *Autumn* | |
| Beginning of autumn (*liqiu*) | Aug 7–9 |
| Limit of heat (*chushu*) | Aug 22–24 |
| White dew (*bailu*) | Sept 7–9 |
| Autumnal equinox (*qiufen*) | Sept 22–24 |
| Cold dew (*hanlu*) | Oct 8–9 |
| Frost's descent (*shuangjiang*) | Oct 23–24 |
| *Winter* | |
| Beginning of winter (*qiufen*) | Nov 7–8 |
| Slight snow (*xiaoxue*) | Nov 22–23 |
| Great snow (*daxue*) | Dec 6–8 |
| Winter solstice (*dongzhi*) | Dec 21–23 |
| Slight cold (*xiaohan*) | Jan 5–7 |
| Great cold (*dahan*) | Jan 20–21 |

is divided into six solar units, subdivided into two groups of three (see Table 7.5).

These solar terms and the entire classification system are utilized all over China. Indeed, they are so common as to appear in any standard dictionary. It is not unusual that Mongol herders use this system to mark time. What is more distinctive in the pastoral villages of Wengniute, however, is the fact that other supplemental systems of marking time that are commonplace in rural China (such as market days, market fairs, holidays, anniversaries, religious festivals, careful attention to the lunar calendar) are either much less conspicuous or do not operate in the same systematic way. For example, there are occasionally special community ceremonies held during the summer (such as *nadaam* fairs and *oboo hui*), but these events are not rigidly scheduled. They may or may not occur in any given year, and last-minute cancellations are not infrequent.

In general, Mongol sensibilities of space and time remain distinctive from industrial norms, but they are slowly evolving toward greater alignment with the standards that intrude from farming and even urban environments. With the advent of the reform era, there are clear signs that changes in the tempo of daily life are beginning to occur. One of the most obvious manifestations of time acceleration is visible in changing labor patterns.

The whole point of land enclosure is to promote more intensive utilization of increasingly scarce productive land. More intensive land use always requires greater inputs of labor, which typically also entail greater inputs of capital (Boserup 1965: 43). Labor intensification necessarily increases the social value of time, thereby accelerating the pace of community life for all activities. In Nasihan, as labor commodification slowly intensifies in association with economic reform and land parcelization, time becomes more precious and the day more regulated. Those who migrate outside the community for work suddenly find themselves earning a wage calculated by daily or even hourly output. Those who sell their labor in the community now also contract themselves out for more precise and shorter increments of duration. For example, they increasingly prefer payment based upon unit of output rather than upon number of days committed to the task. Time has become more valuable, so residents have adopted a more discriminating system of compensation. Not long ago, herders accepted wages for their skills based upon each unit of livestock managed per year. As the *lianhu* dissolve and more families take to managing their own herds, however, professional herders now tend to calculate their fees on a shorter, seasonal basis.

Accelerating land degradation and the use of fence-wire both further contribute to the local intensification of time awareness in at least two important respects. First, there is the principle of pasture reconstruction, which explicitly invokes the advantageous manipulation of time management. An enclosure is a synthetic space deliberately constructed to speed up and slow down particular processes of nature. As Mircea Eliade once observed, such interventions on the landscape signify nothing less than humanity taking upon itself the role of time (cited in Jackson 1984: 8). Even without the influence of labor intensification, Mongols who enclose rangeland are effectively accelerating community experiences of time simply by speeding the tempo of private pasture regeneration and public pasture degradation.

The recent penetration of investment capital constitutes the second significant channel by which enclosures explicitly impose new orientations of time on the community. In association with pasture reconstruction, independent households are beginning to apply for bank loans at interest. This vehicle directly exposes them to unfamiliar monetary time values. Even though the number of recipients remains small, the growing desire for working capital itself generates new thinking about the cash value of time.

Of course, all these influences are reinforced by the broader context of expanding commercialization. Residents increasingly find their interests and lives connected to urban centers far beyond township borders. A handful of community pioneers waits for public bus service beside the dusty gravel road with increasing frequency and growing impatience. Public bus service has recently expanded to two routes, and a private entrepreneur from Wudan now provides a more expensive "express" service every other day. Television has also become increasingly prevalent in the village since it was introduced over the last decade. Despite limited programming, the prospects of nighttime entertainment for many homes has made the hour and the minute a socially meaningful increment for the first time.

Indigenous space-time orientations are slowly evolving into more bounded and rigid increments. Harvey (1990) refers to this phenomenon as "time-space compression" and considers it one of the many consequences of economic modernization for traditional societies. In Inner Mongolia, as elsewhere around the world, alien life-world standards increasingly penetrate the consciousness of indigenous populations as they come into greater contact with manifestations of the capitalist world system.

## Personal Accounts

Before concluding this chapter, I would like to introduce three different householders in order to illustrate in a more specific way how private enclosures are redefining access to community resources and creating new economic disparities within the township. All three live in Wulanaodu, but they belong to different kinship groups and represent different wealth strata. They all belong to households with whom I shared a great deal of time and many intimate experiences while conducting my research. I was a guest in each of their homes on a regular basis and exchanged gifts or favors with them frequently. Indeed, my feelings for them make it difficult to limit my comments to a brief summary of their life history and current production circumstances.

FIG. 7.5. Dabagan sitting on mule-litter with friend

*Dabagan.* Of all the remarkable people I met in Nasihan, none had a more interesting personal history than this seventy-three-year-old gentleman (Figure 7.5). In many regards he is an atypical resident and a poor choice to represent the village's lower income stratum. But his life story is so compelling and his poverty so undeserving that it does help to clarify many of the social issues raised in this book.

Dabagan was born in Liaoning Province in 1921 under the sign of the dog. At the age of four, his father died, and his stepfather drank so heavily and beat him so often he ran away from home at the age of thirteen. He survived by begging for food and accepting any menial work he could get. He took a job sweeping floors at a Japanese military academy in Tongliao until an officer took notice of his intelligence and sent him to school. He was later admitted to a high school in Hailar, one of the best in all of Mongolia, and received a Japanese education. At the age of nineteen, he was drafted to serve the Japanese military in their occupation of Manchuria, but he escaped on a train along with six classmates who fled westward to join the Mongolian Independence Movement led by Prince Demchukdonggrub (*De Wang*) in Hohhot. He was put to work as a police detective and military translator in Baotou and Zhangjiakou, since he could speak Japanese, Chinese, and Mongolian.

When Japan surrendered in 1945, Dabagan was only twenty-four years old and decided to join the Nationalist Army under the command of General Hai Fulong. He was a member of one of the first platoons to penetrate the Northeast front lines, engaging the famous Communist troops of the Eighth Route Army (*Balujun*). Dabagan reports that because the Nationalists were so suspicious of Mongolian soldiers at the time, and because of Ulanfu's great influence over Mongolian soldiers, his immediate commander organized an insurrection and persuaded them all to seek refuge among the Communists on the night before the Communist siege of Datong began. He then joined in 1947 the Seventeenth Division of the Inner Mongolian Communist Liberation Army and served as section chief of logistical supplies in a guerrilla outfit. A few months later he was transferred to Daqingshan for revolutionary indoctrination, which proved to be a life-saving alibi when he was imprisoned by the Communists on charges of treason after a group of soldiers in Datong abdicated to the Nationalist Army. He was released in 1948 and returned to his ancestral home in Wengniute with no further interest in military affairs. The countryside was in such chaos that he was happy to take a job as a cattle herder offered by an old acquaintance.

Once the war ended, his intelligence and literacy again made him attractive to district Party officials. They dispatched a large white horse to the field where he was grazing his herd and conscripted him to join their administration. He refused to accept the duties of office secretary and opted instead for a position as teacher and principal of a Mongolian school. In 1949, he was elected to several educational committees, which required him to travel into Chifeng City. En route, he often stopped in Nasihan and there he eventually met and married his wife. He earned a decent salary for a few years, but once the political campaigns associated with land reform began to sweep the countryside, his previous experiences became terrible liabilities. Because he had once served as a Japanese translator and a member of the Nationalist Army, he was the first target of local struggle meetings and physical abuse throughout the 1950s and 1960s. In 1958, he was forced to undergo correctional discipline for several years in a prisonlike institution in Chifeng City. Since then, he has been stripped of all social standing, beaten severely on many occasions, and ridiculed throughout the community ever since with a nefarious nickname: "the old spy" (*lao tewu*).

During the collective era, his only daughter developed leukemia and died at the age of fifteen after she had consumed all of his resources on medicine. Then the Cultural Revolution commenced and Dabagan was accused of leading Gengden, the "founding father" of Wulanaodu and current party secretary of the commune, into the ranks of the *Neirendang*. For eight full years, Dabagan was tormented with false charges and brutal interrogations. His wife was also injured and even blinded by Red Guards because her father had been labeled a "rich peasant" *(funeng)* during earlier land reforms. Since the Cultural Revolution, Dabagan has never been able to collect any of the financial compensation officially due him for the losses he suffered during that era. Quite the contrary, he has spent much of his own money and slaughtered many head of cattle to cover the expenses of trying to get his name cleared of any wrongdoing during the turbulent years of the Chinese Revolution. He craves exoneration both for vindication and for protection against the next political campaign that comes his way. His persistent efforts over many years did bring some success, as he eventually received a medal of meritorious service from a government office in Hohhot, which essentially absolves him of culpability for the complicated political circumstances of his youth.

Dabagan is one of the most intelligent and cosmopolitan residents of Nasihan, yet he lives in almost abject poverty. He inhabits a tiny mud hut that appears on the verge of collapse. His only possessions include a chest of drawers, a tea table, some cookware, some bedding, and a radio that he plays all the time to keep up with current events. In 1994, he owned only two cows, two sheep, nine goats, two pigs, four chickens, and a dog. He keeps his livestock in a pen near his house through the winter, but sends them to pasture in the summer under the care of his wife's brother, an experienced but aged herder who takes most of Dabagan's hay for compensation. Dabagan controls only one plot of enclosed land about a hectare in size, which he received upon decollectivization. The fertile plot was formerly a garden area that he had managed for his production team since 1975. It was already fenced with cheap wire mesh when he acquired it, and he continues to cultivate the land with carrots, corn, soybeans, and turnips. He estimates that he now earns about 400 yuan per year selling surplus vegetables, 550 yuan selling goat hair, and he draws 740 yuan in fixed government pension. On the other hand, he spends about 500 yuan each year on grain and noodles, and another 500 each on liquor and medicine. His most recent tax burden totaled 138 yuan.

In the last few weeks of my stay in the village, I watched him serve his bedridden and dying wife faithfully and with tender affection. He planned, upon her death, to undertake an extensive religious pilgrimage from which he would not return. I salute him and mourn his passing; he was an invaluable inspiration to me and my research.

*Zhamusu.* Born in 1953, Zhamusu came to Wulanaodu at age six when his herding household settled there to join the collective. His childhood was marked by poverty and hunger as his parents tried to provide for five sons and five daughters. Indeed, at one point they became so destitute that they gave away their youngest son to be raised by a childless widow in the village. Zhamusu went to local schools from age eight to age sixteen, at which time they closed as a result of the Cultural Revolution. He participated actively in the Red Guard struggle meetings and accepted their ideology. At eighteen, he resumed high school in an adjacent township, then returned to Wulanaodu after graduation. He tried to join the army, but the Viet Nam war had just ended and military conscription was suspended. A production brigade charged him with conducting the first land survey for the entire township. He was subsequently asked to serve as a schoolteacher, since his educational level was higher than that of most other residents. At twenty-three, he joined the army and served five years in the northern district of Xilingele. He was delighted to be accepted into the Communist Party at twenty-six, after passing a thorough background check and writing lengthy essays about how he planned to contribute to the development of socialism in China.

At his request, he left the army in 1980 and returned home to find his family in shambles. His mother had died after a stroke at the age of forty-eight, and his father had become too old to shoulder the burden of labor. As the oldest son, Zhamusu felt responsible for organizing the recovery of his household and to oversee the affairs of his three adolescent brothers. He first went hunting and killed several hundred rabbits and a dozen foxes to obtain some working capital. He also earned some money repairing electronic instruments, a skill he acquired in the military. He then purchased some livestock and directed his brothers in their herding responsibilities. Before long, Zhamusu accepted a salaried position as a township-level cadre, charged with directing the office of cultural ministry. It was a moment of triumph for him and his family.

During this phase of his life, he also married a complete stranger through a low-budget matchmaker. The union lasted only two years before he sued for divorce. The woman resisted until Zhamusu threatened her with the warning that "he could not be responsible if she had some accident in the future." He also relinquished to her the lion's share of their possessions, including a male son. Four months later he went to

FIG. 7.6. Zhamusu and brother harvesting hay

Wudan in search of another wife. He found someone with a good temperament and applied for marriage even though she was only seventeen (under age by one year). He used his position in government to get around the law, but the deceit was discovered and he was summoned to account for his actions. When he explained that he feared she would find another suitor rather than wait a full year to marry a man fifteen years her elder, the ruling official was sympathetic and let him off with a warning to make his marriage a model of success for the Party. The marriage has since produced three healthy sons, and Zhamusu was fined 500 yuan for exceeding the birth quota. He complains that the family planning policy changed after his wife was already pregnant, so he was caught by circumstances beyond his control.

Zhamusu has brought his own household and that of his father into the middle class of the village. He initiated family division (*fenjia*) with his father in 1992 when he purchased a house, but still works closely with his father and brothers to pursue their mutual prosperity (see Figure 7.6). The youngest brother is still single and leaves Nasihan every spring to sell his labor in urban markets. He has worked in brick factories in Hailar and built houses in Shenyang, bringing cash back to the family at the end of autumn. (Migrant workers earn about 240 yuan per month.)

He is proud of his contribution to the household and reports that the Han love to hire Mongols because they work so hard. Another brother is still single after watching four previous marriage engagements collapse over last-minute economic disagreements. He has another prospect on the line with less negotiation leverage because her hair color is light and Mongols apparently prefer dark-haired women. This brother lives out on the range, caring for his father and tending to the entire family's live-stock. The third brother is crippled and lives in the residential center with a wife and small daughter beside Zhamusu.

In 1994, Zhamusu owned five cows, twenty-eight goats, and one mule. He shares rangeland resources with his brothers and has gradually developed their pasture holdings into three separate enclosures of about 13 hectares each. He enclosed the first portion in 1986 with fence-wire obtained through his influence in the local government. He enclosed the second portion in 1989 and the third in 1993 with fence-wire obtained by selling off sheep and other livestock. He estimates that he made about 3,000 yuan in 1994 selling livestock produce, and another 1,500 yuan from his cadre salary. He has also initiated some sideline ventures, earn-ing 1,500 yuan for electronic repair and photography services. Since he travels so much with his job, he plans to open a small business in petty trade, selling cigarettes and other small commodities out of his home. He estimates that he spends about 700 yuan per year on grain and noodles, 450 yuan on liquor, and about 300 yuan on medicine for his anemic wife. His capital assets total about 5,000 yuan and his most recent tax liabili-ties totaled 238 yuan.

Zhamusu is determined to prepare a bright future for his three boys. As a member of the local government, he is careful to express support for the enclosure policies, and he looks with pride and satisfaction upon his own economic advances in recent years. Yet he struggles with the am-bivalence and insecurities of the emergent middle class. In more reflective moments he acknowledges the bad relations that fencing has caused be-tween neighbors and explicitly identifies with the weak when he raises his voice and declares, "in the future, when a Wang puts up an enclosure on our land, my sons will be able to push the fence back on them." Zhamusu seems quick to adopt a combative tone, perhaps because he is still emotionally sensitive and professionally vulnerable to the memories of his impoverished childhood. I witnessed, for example, one incident in which the village chief of Wulanaodu (a member of the Wang clan) be-came drunk at a festival and publicly ridiculed Zhamusu as an inferior man who did not know his proper place. The antagonist jeered, "How is it that you, a township-level cadre, must walk the desert on your own two feet while I ride a powerful horse? In the future, you will be back at my gate to beg for food." Such talk is not idle chatter. Considering the dramatic environmental and social transformations going on around them, it is the constant fear of just such a scenario that looms in the mind of every local resident.

*Haili.* Born in Nasihan in 1955 to an exiled medical doctor, Haili was named after a beloved ancestral town to which the displaced household could not return. Haili's father had studied medicine in Chifeng City and took his first job in Hongshan in a hospital for genital and reproductive diseases. During the "Hundred Flowers" era when Mao invited critical commentary from intellectuals, he wrote an essay for the newspaper that offended some officials, and he eventually became a victim of the subsequent purge. He was fired and then transferred (*xiafang*) to the countryside to learn humility from the simple folk of Wulanaodu. For the rest of his professional career, he ran the township's medical clinic, but he suffered physical abuse and imprisonment during the Cultural Revolution. He raised his daughter to be a nurse in the clinic, and directed his youngest son into medicine as well, but he allowed his oldest son, Haili, to pursue the practical business of herding. Haili enjoyed pastoral activities more than books and learned the skills of animal husbandry while working for the collective.

Haili never moved away from Wulanaodu, having married a local woman. They have two children, a boy and a girl, and they now share a large brick house with his parents. Haili is not noteworthy for an interesting biography, but for his financial success during the reform era. Starting with only an average-size herd after decollectivization, but an excellent pasture allotment, he has skillfully managed the growth of his livestock and gradually become an active entrepreneur and major economic player in the village. Though Haili does not belong to the Wang clan, he has become an investment partner with some influential Wang family members. This was facilitated by the fact that Haili's older sister married into the Wang clan years ago. It also reflects the fact that Haili is witty in conversation and shrewd in business, and highly popular among his cohorts. In any case, he is not representative of many elite households who acquired an early advantage primarily through Party influence or privileged bank loans.

He did, however, play his cards well with the research station scien-

FIG. 7.7. Haili capturing cow in front of his opulent winter hay supply

tists, who signed him on as a "model" citizen to demonstrate advanced
production techniques to other local residents. In exchange for this important ideological service to the visiting scientists, Haili enjoyed special
favors such as the use of irrigation equipment, production tutelage, enhanced grazing and mowing rights on research station property, transportation assistance, and access to fence-wire. I have no way to quantify
how much assistance he received or how much of an edge it gave him
over the years. I can, however, report on how well he has prospered.

In 1994, Haili maintained 21 cattle, 5 sheep, 101 goats, and 2 horses.
He now controls about 100 hectares of land. Most of it is out on the
range, but seven prominent hectares lie in the residential center adjacent
to the community reserve hayfield. He first enclosed his land base in
1988 in cooperation with a herding specialist who looks after the animals
on a daily basis. They graze all their animals outside the enclosure until
nothing is left to eat. Haili is now an active employer in the region, hiring other men to produce hay, card goat hair, and weave fence-wire—not
because he does not do the work, but because he has too many resources
to control with his limited household labor. He now estimates his capital
assets to be worth about 22,000 yuan, with a total annual income of
13,950 yuan. With operating expenses on the order of 5,500 yuan, his

net income approaches 8,450 yuan. His most recent tax liabilities totaled 850 yuan.

Haili gained most of his wealth in just the last few years by engaging in direct marketing. He began buying livestock from his neighbors, only to transport and sell them to the regional stockyards at more favorable prices. He started out small but now does this on a large scale, trading dozens of cattle every autumn (see Figure 7.7). He does the same thing with goat hair and sheep wool in the spring. Recently, he has begun to expand into the trade of other commodities as well. He buys bricks, coal, and grain in bulk from urban suppliers and then sells them to residents of Nasihan at incremental profits. So far, he has plowed most of his money back into pastoral investments (livestock, fence-wire, fertilizer, hay supplies), but he also tries to expand his volume of trade in sideline commodities every year. Lately, he has even begun to sell fence-wire. It was Haili who first found a source of supply for the recycled wire mesh that has become a popular alternative to more expensive wire. When asked about those neighbors who cannot afford to enclose their land, he shakes his head and replies, "those with no money to buy fence-wire are finished . . . they have lost their production base." When asked who took the land from them, he defensively blames the many households in the village even wealthier than he.

# EIGHT

# Landscape and Identity

Ethnic Mongol herders in China have traditionally enjoyed a distinctive home environment that contrasts significantly with that of the agricultural Han. Furthermore, their senses of space and of place are tied intimately to a sense of self, so that the recent disruptions in spatial, temporal, ecological, and social organization only feed community disorientation. Settled but not yet sedentarized, the herding community of Nasihan generally experiences the enclosure movement in the context of a deepening crisis of identity.

In areas where pastoralism is still a way of life, but where the traditional environment of pastoralism is under assault, the landscape itself has become a potent medium of symbolic communication between residents as they actively renegotiate their individual and collective identities. Mobile pastoralism, after all, is not just a subsistence strategy, but a fundamental relationship with the world.

Considering the historical role of Wulanaodu as a "model" grassland community in the forward progress of Maoist socialism, interpreting the landscape could hardly be a trivial matter in the region, either for residents or the state. Once a symbolic center for the promise of all Chinese pastoralism, the physical environment of the famous "red star" brigade is highly charged with political inference. Local landscape is thus predisposed to play an important role in the social conflicts that have emerged with decollectivization. Given the high social stakes involved in the enclosure movement, three features of the landscape have become especially meaningful: grass, sand, and fence. The people of Nasihan clearly pay attention to them in thinking about themselves and their social relations.

In reality, these three features are but separate aspects of a single concern. They all reference the same transformation precipitated by the proliferation of household enclosures. Generally speaking, fence has become the new dividing line between sand and grass. Where grass is absent, there is sand, and vice versa. This is a familiar reality. However, the former patchwork pattern at a landscape scale has been increasingly restructured in tandem with the appearance of barbed-wire fencing. Where sand and grass once mingled intermittently across the range, now they separate more neatly into discrete concentrations that follow the shadow of private fence lines. Increasingly, these three features constitute a single constellation signaling the naturalization of inegalitarian ideology. By "naturalization," I mean that environmental stratification corresponds with social stratification so that both processes mutually reinforce each other in a manner that seems perfectly natural—elite residents control the most fence-wire and the most grass.

Residents often shared with me some introspective reflections about their changing life-world. They consider themselves to be transforming into a rather strange breed of people—in the words of one articulate man, "not quite farmer and not quite herder, not quite Mongol and not quite Han, not quite traditional and not quite modern." They understand themselves to be under transition, and they sense that many daily activities that now occupy most of their time will not fit their lives in the near future. They know their production capacities are quite limited both as herders and as farmers, but they do not have the wherewithal to specialize exclusively in either endeavor. They can see the inevitability of a less extensive grazing system, but they cannot see how to make it profitable with the labor, capital, and land at their disposal. They know that more transformation is coming, but they do not know what it will entail or how they will survive it. Most are not optimistic.

## Settled, Not Sedentary

As Mongols have increasingly practiced a more settled version of pastoralism over the twentieth century, herding communities have made obvious adjustments in their traditional lifestyle and cultural forms. For example, throughout Wengniute, yurts gave way to mud-brick homes several generations ago, and cultivated fields have come to dissect many pastures. Though the once totally dominant experience of mobility has been significantly modified, the daily practice of free movement in open space has persisted throughout most of rural Inner Mongolia right up to the present day. In areas like Wulanaodu, where decollectivization prompted the rapid proliferation of household enclosures, pastoral communities have struggled to maintain a distinctive, though evolving, spatiality that still defines social practice and sets herders apart from Chinese farmers.

In contemporary Wengniute, mud walls, which are ubiquitous throughout the western farming communities, abruptly vanish as one travels eastward and passes into pastoral townships. Of course, fences now crisscross the wide open pastures, but these areas still communicate their own sense of vernacular space. The built structure on the landscape, though settled, is still not sedentary. For example, the fences still scream impermanence in their every aspect. At times, whole sections of fencing lean or even lie trampled on the ground as livestock wander serenely across. Wire is often broken or missing.[1]

To a Western observer it first may seem that many residents do not know how to set a post, run a fence line, or use their equipment properly. But then alternative explanations arise: Perhaps their hearts are simply not in the work—or more likely, perhaps indigenous ideas of fencing just have greater tolerance for permeability than outsiders think appropriate. In my experiences as a fellow pedestrian with village residents, I observed that natives expect to cross any fence in their line of motion by pushing the wires below their waist while stepping over. It is too burdensome (and demeaning) to search for the gate. They are quite put out when fence-wires are insufficiently slack to accommodate their desire to pass. As they see it, a fence should corral livestock, not humans. From this perspective, the indigenous fence actually reinforces this abstract ideal of mobility rather than constrains it. This abstraction is reinforced in another way as well. Fences act as a mechanism by which grass may be transferred from that unenclosed spot over there to this private pasture over here. Residents watch this happen all the time.

Beyond the fences, the ideal of mobility continues. The majority of houses are still made of mud, with an estimated lifespan of only ten years. They are constantly mended and rebuilt. Abandoned and dilapidated structures are a common sight in the area; they often stand just a few meters from newly constructed replacement dwellings (which can be erected within a week). Indeed, the entire village landscape remains highly mutable, with structural changes that imply more than just tinkering in the yard. A windbreak goes up here and is torn down again after two days; a cow shed disappears overnight; a prominent tree is removed and the village road displaced in the flurry of a morning of work. The most dramatic changes around the homestead occur in preparation for winter. For example, temporary wooden shelters for livestock are built or expanded, ensilage depots are reconstructed, heavy screens of woven willow twigs are strategically erected to channel windflow, and dung and wood fuel is stockpiled high beside the house. Adjustments to these seasonal structures go on all winter long before they are removed again at the beginning of spring and a different set of transitory features predominate. To an outsider grasping for orienting landmarks, the whole setting has a surreal volatile quality.

Along with the transient landscape, the residents themselves are involved in a parade of movement that goes on year round. They frequently move their animals, their children, and their supplies back and forth between village center and distant pastures. They constantly scavenge for willow twigs and cow dung to burn in their homes for heat and cooking fuel. They peddle their hides and surplus produce among neighbors. They routinely (several times a month) journey by mule-litter at least twenty kilometers round-trip into the central township village to purchase home supplies and foodstuffs. During autumn months they are especially mobile. Horse-drawn carts circulate everywhere, delivering family labor to cut hay in the reserve meadows, then transporting the tied bundles back home. I observed that during the month of September, some families made more than thirty round-trip cart deliveries between their homes and the hayfields some four kilometers away. After that, all winter long, the hay must be transported again to the livestock stationed far out on the range. This arrangement is not exactly nomadic, but neither is it settled. Rather than move their animals to fresh seasonal pastures as Mongols have done for centuries, Nasihan herders now essentially move the grass to the animals. But they still move.

Table 8.1 provides a seasonal account of the "typical" movements of one household through a single year—that of Zhamusu, his wife, and three children (see Chapter 7), who live within the residential center of Wulanaodu and share enclosed pasturage on the range with Zhamusu's two brothers. Obviously, the distances reported here are only rough estimates and might vary significantly from household to household, depending upon individual circumstances. I selected this family because I know their circumstances more intimately than most (I helped them cut and deliver hay), and because their relative proximity to the collective hayfields makes their numbers a *minimum* standard of measure for the kind of activities that all active pastoral households undertake. It is worth emphasizing four points about these distances: (1) they are covered in the course of activities generally involving the entire family; (2) they consti-

TABLE 8.1
*Annual Household Mobility Peaks Related to Production*

| Season | Activity | Distance involved (round-trip estimates): |
|---|---|---|
| September 1–15 | cutting hay | journey to field (15 x 6 km): 90 km |
| September 15–30 | storing hay | delivery of 24 cartloads home: 144 km |
| October 1–15 | harvesting crops | journey to field (3 x 4 km): 12 km |
| November 1–10 | weatherproofing | journey to mud pits (4 x 4 km): 16 km |
| November 10–20 | hunting | irregular tracking (3 x 20 km): 60 km |
| November 21–25 | cutting willow | journey to meadow (3 x 10 km): 30 km |
| November–January | transporting hay to pasture | delivery of 24 cartloads (14 km): 336 km |
| November–April | foraging fuel | daily walking circuit (2 km/day): 360 km |
| April 20–30 | goat shearing | journey to pasture (2 x 14 km): 28 km |
| May 10–15 | plowing/seeding | journey to field (5 x 4 km): 20 km |
| June 15–20 | sheep shearing and dipping | journey to pasture (2 x 14 km): 28 km |

tute mobility in excess of normal routines associated with livestock herding, trade, moving supplies, and general sociability; (3) much of the mobility results from a dual-residence animal-husbandry management system that actually functions much like the traditional division between winter and summer camps; and (4) the terrain is characterized by deep sand that significantly complicates movement.

## Persistent Perceptual Thresholds

The traditional ecological attitudes of Mongol herdsmen still exist in Nasihan and manifest themselves in subtle ways. Salient discrepancies emerged when I began to contrast the language of Han scientists at the research station against casual comments elicited from local residents. Contrasting use of the Chinese term *huang* (waste) provides one clear example. I have already discussed (in Chapter 4) the negative associations of that term within the general context of a national frontier discourse. The itinerant Han scientists in Nasihan consciously sustain that discourse in their daily work. Indeed, the director of the research station frequently expresses the station's operational goals in the region by repeating a simple slogan: We must turn "yellow" (*huang*) into "green" (*lu*)—that is, turn sand into vegetation. (The slogan conveys a pun because the Chinese phoneme *huang* can mean both "yellow" and "wasteland.") From his ethnic and professional perspective, the patchwork of dune sand in the local environment is both aesthetically displeasing and agriculturally useless. It is best converted into fields of intensive fodder cultivation and tree farming.

In contrast, local Mongols use the term *huang* more positively. They speak of *huang* as yellow-tinted "living sand" and contrast it with white "dead sand" (*bai shazi*). They perceive that yellow sand can sustain vegetation, while white sand will not. Therefore, only white sand deserves to be considered infertile, though it also has a certain utility (as I discuss later). Yellow sand, to their way of thinking, is "good sand" (*hao shazi*) because it remains "alive" (*huo*) with potential.

I became intrigued with exploring the possibilities for determining more distinctive perceptual thresholds. I decided to use photographic prompts as a means to standardize and quantify resident responses to questions about landscape preference. After living in the township for several months, I worked in collaboration with a few local informants to photograph a variety of typical landscape scenes. I then selected twelve photos that represented a broad spectrum of variation. Some photos portrayed reserve hay meadows with dense grass growth, others portrayed images from unenclosed range with only moderate vegetation, while others portrayed images dominated by sand. My plan was to ask respondents to arrange the photos in rank order according to their own perceptions of immediate land quality for the purposes of livestock grazing. After that, I would ask them to refer to the photos to answer other questions pertaining to qualities of resilience and aesthetics. By way of comparison, I also arranged to interview the regular staff members among the Han scientific community at the research station.

There are significant methodological problems associated with the use of photographs, but in my field setting the benefits far outweighed the liabilities. In such a remote and impoverished field setting, my presence was initially intimidating to many residents. Most of them had never even met a foreigner, so the idea of hosting one in their homes was too overwhelming without the aid of some mechanism to break the ice. The photographs not only captured their immediate interest but also firmly established their roles as experts. Furthermore, the photographs were a medium that allowed all of us to get beyond the initial problem of finding precise terminology to explore and discuss the meaningful features of their physical environment.

## PERCEPTIONS OF ENCLOSED LAND

Local residents confirmed my own analysis that household enclosures are dramatically restructuring the ecological environment. There was a strong positive correlation between the perception of abundant fodder

TABLE 8.2
*Frequency of Perceived Enclosure of Land*
*by Relative Rank Order Position*

| Relative rank order | "Land is enclosed"[a] | "Land is unenclosed[b] |
|---|---|---|
| 1 | 105 | 7 |
| 2 | 103 | 9 |
| 3 | 99 | 13 |
| 4 | 98 | 14 |
| 5 | 88 | 24 |
| 6 | 74 | 38 |
| 7 | 50 | 62 |
| 8 | 45 | 67 |
| 9 | 32 | 80 |
| 10 | 24 | 88 |
| 11 | 8 | 104 |
| 12 | 3 | 109 |

NOTE: Total respondents = 112.
[a]No. of respondents perceiving that land is enclosed.
[b]No. of respondents perceiving that land is unenclosed.

and the expectation that a particular parcel of land was enclosed. During the survey, I asked each respondent to separate all pictured landscapes into two piles: one for those believed to be inside an enclosure, and one for those believed to be outside all fenced enclosures. I instructed each respondent to make their decisions based solely upon the visual cues of each photograph, rather than upon any presumed familiarity with the territory represented in the picture. The responses for each rank category are summarized in Table 8.2.

As the relative rank of fodder quality increased, so did the expectation that the landscape had to be enclosed. Given the context of competitive enclosing during the 1990s, this expectation was based on commonsense daily experience. Respondents generally displayed no hesitation separating the piles because they followed a simple guideline: if the grass patches were dense, the land had to be enclosed. This clearly demonstrates that residents perceive household enclosures to be disrupting the indigenous patchwork of the range by dividing it into more discrete concentrations of vegetation and sand. This is a troubling rearrangement of rangeland resources for most residents, who continue to prefer a biologically diverse regional landscape in which to herd their animals.

PREFERENCE FOR LANDSCAPE DIVERSITY

During the survey, resident herders consistently indicated that they placed a high value on landscape diversity. Height and density of grass was seldom their only consideration in evaluating rangeland preferences, as it was when I queried personnel at the research station. The presence of trees, hill slopes, and even patches of sand were all deemed important components of a desirable grazing environment. Respondents typically used the photos to explain that livestock grazing was a mobile activity that thrived on landscape variability. Grass is good to eat, they conceded, but animals also need browse matter, moisture, shade, protection from wind, and exposure to many kinds of forage, both within a single season and in different quantities throughout the year. To Mongol herders, the relative value of a given pasture will always depend upon the season of use.

Traditionally, for example, a suitable winter pasture could do without water (which was provided by snow) but absolutely required a good wind break. Suitable spring pastures required position on the southern slopes of a hill, where snow melts and grass grows the quickest. Summer pastures required access to water, grass, and soda licks, while autumn pastures primarily required particular grasses that promote lactation and fat buildup (Szynciewkicz 1982: 22).

Landscape diversity also conditions the quality of grazing within each season. Especially in the summer, when an abundant variety of grass is necessary to assure good lactation and thick fleece growth, ecological heterogeneity of forage is critical. A study by Fernandez-Gimenez (1995: 10) reported that landscape diversity (both between and within seasons) is consistently valued among herders all across Mongolia. My interviews confirm the persistence of these attitudes among contemporary Mongol herders in China, despite government efforts to sedentarize the population.

Of course, water and grass constitute the most essential grassland resources. Indeed, grass is actually more of a multidimensional resource to herding communities than often assumed. In addition to its obvious utility as browse and fodder, it has value in many other ways. Stored as hay, grass constitutes a highly liquid asset that can be traded or sold just as any other crop. Also, many grass species in Inner Mongolia are considered medicinal to humans, so they are scavenged both for domestic use and for exchange value. Other species are highly marketable because of ascribed symbolic value. For example, one variety enjoys a popular name whose characters, when pronounced, sound just like the Chinese term for "strike it rich" (*facai*). Urban restaurants buy the grass for use in a special soup that many patrons consider auspicious. (This consumer demand

FIG. 8.1. The "best" landscape

FIG. 8.2. The "worst" landscape

also contributes to local land degradation processes, as sod is uprooted for direct market exchange.) Grass also functions locally as an ecological monitor, for it helps both to check erosion and to indicate soil fertility. Furthermore, grass roots are highly prized as construction material. Some of the poorest residents build houses out of sod, or thatch a roof with it. Finally, some grass species provide important craft materials, such as broom straw for making baskets and various hand implements.

Grass is probably more diversely important to pastoral peoples than Western readers commonly realize, but there are many other grassland resources that also contribute in surprising ways to the viability of pastoral production systems. These include environmental components such as solar energy, wind energy, timber, minerals, and even dune sand. Although water and grass are basic, the other resources also play important roles in production and household economic strategy. Dune sand provides the most interesting case in point.

RANGELAND AS RESILIENT AND AESTHETIC

Resident herders were surprisingly optimistic about the presence of dune sand and the resiliency of the land. Many members of the community, including some of the most educated and elite, expressed the opinion that even landscapes totally dominated by moving dunes can be restored to "full productivity" (*huifu shengchan nengli*) within a mere three years' time.

To be specific, during one survey, I asked 130 household managers across four different villages to estimate the number of years required before the least desirable landscape represented in my photos could recover to the same approximate level of productivity as the most desirable landscape, once enclosed and left fallow. The photos under comparison necessarily varied, depending upon each respondent's prior selection of "best" and "worst" grazing landscapes. One photo was fairly consistently chosen as "worst" (106 out of 130 valid cases), while four different photos competed closely for the position of "best" (together comprising 112 out of 130 valid cases). All four most popular candidates for the "best" landscape depicted a stable terrain almost fully covered by green vegetation. In the majority of cases, the respondents were comparing Figure 8.1 (best) against Figure 8.2 (worst).[2]

The responses to the question of resiliency are listed by frequency in Table 8.3. Among village respondents, only two said that the worst landscape would never recover to the level of best, and three others said that

TABLE 8.3
*Resident Estimation of Years Required to
Restore Land Productivity*

| Years | Frequency | Percent | Cumulative percent |
|---|---|---|---|
| 2 | 11 | 8.5% | 8.5% |
| 3 | 33 | 25.4 | 33.9 |
| 4 | 15 | 11.5 | 45.4 |
| 5 | 25 | 19.2 | 64.6 |
| 6 | 5 | 3.8 | 68.4 |
| 7 | 8 | 6.2 | 74.6 |
| 8 | 1 | 0.8 | 75.4 |
| 9 | 1 | 0.8 | 76.2 |
| 10 | 26 | 20.0 | 96.2 |
| 15 | 1 | 0.8 | 97.0 |
| 20+ | 4 | 3.0 | 100.0 |
| TOTAL | 130 | 100.0% | 100.0% |

SOURCE: Interview data.

TABLE 8.4
*Scientist Estimation of Years Required to
Restore Land Productivity*

| Years | Frequency | Percent | Cumulative percent |
|---|---|---|---|
| 5 | 1 | 12.5% | 12.5% |
| 8 | 2 | 25.0 | 37.5 |
| 15 | 1 | 12.5 | 50.0 |
| 18 | 1 | 12.5 | 62.5 |
| 20+ | 3 | 37.5 | 100.0 |
| TOTAL | 8 | 100.0% | 100.0% |

SOURCE: Interview data.

it would take more than fifteen years. The response with the greatest frequency, however, was "three years" (33 out of 130). The average estimate for all village respondents was 5.9 years. In contrast, five out of eight willing respondents among the Han scientists at the research station said that the same sand covered landscape would require more than fifteen years to restore a stable vegetation cover. The average estimate of recovery time from the scientific community was fourteen years, or nearly two and one-half times longer than the view from native perceptions (see Table 8.4). According to the scientists' published research, enclosed dune sand only begins to show marked increase in aboveground biomass after at least seven years fallow (Kou and Xue 1990: 6–7). The senior director of the station argued that their experimental plots had not fully recovered even after twenty-five years of protection.[3]

These contrasting responses indicate that local residents have distinctive perceptual thresholds with regard to both the resiliency of the land and the minimal appearance of a landscape that is worthy of the term "full productivity." I had the opportunity to advance follow-up questions during interviews, and so was able to discern that most respondents did not literally believe that the worst plot of land would necessarily replicate the best in three years' time. (For example, they admitted the grass species and vegetative cover might not be the same.) Rather, their responses indicate that their notion of fully productive land is highly tolerant to the presence of sand. Consistent with their value of landscape diversity, a pasture characterized by sufficient patchiness of multiple resources will qualify in their minds as "productive land." The restored pasture would be comparable in use value, regardless of whether it was technically comparable in all aspects of physical appearance. This interpretation of the data helps to explain why so many residents express annoyance that the research station continues to hold so much land in reserve, even after it has laid fallow some twenty years. The Han scientists perceive that the land has still not recovered, but Mongol herders perceive only a capricious hoarding of community resources.

Besides these optimistic attitudes toward land resiliency, the survey also revealed a surprising tolerance for dune sand at a landscape scale. Although the four photos ranked best by the villagers as a whole all exhibited a relatively lush ground cover of vegetation, many individuals actually favored sand-conspicuous landscapes. For example, in twenty-five cases (19 percent), respondents ranked Figure 8.3 superior to Figure 8.4. The same photograph was even preferred over Figure 8.1 (the "best") in 19 cases. I was especially surprised when twenty-three respondents (18 percent) ranked Figure 8.5 superior to Figure 8.4. Figure 8.5 was also preferred over Figure 8.1 in twelve cases.[4]

Whereas Han Chinese are culturally inclined to view a patchy desert-steppe environment as barren and desolate, a surprising number of local Mongol herders tend to view it not only as "alive" but also as aesthetically pleasing. For example, when asked to select the photo that depicted the most beautiful landscape, 37 out of 126 respondents (29 percent) selected a photo that was not among the top four ranked in terms of productivity. Instead of selecting among the most "green" landscapes, signif-

FIG. 8.3. "Diverse" landscape

FIG. 8.4. "Homogenous" field of grass

FIG. 8.5. "Heavily grazed" landscape

icant numbers of the population selected photos in which exposed sand soil or dune sand was again a conspicuous feature of the terrain. Even when sand-conspicuous terrain was not chosen as the most beautiful landscape, respondents typically did not dismiss them out of hand as did the Han, nor did they display any overt revulsion.

## The Economic Utility of Sand

Many herders also appreciate sand for utilitarian reasons. First, residents assert that sand provides a good habitat for livestock. It keeps them dry and hygienic, with a yielding terrain that is neither too abrasive nor too slippery. Second, residents say that dune sand helps to regulate livestock body temperature, keeping the herds warm in the winter and cool in the summer. A poor conductor of heat, sand is valued less for thermal properties than for the shelter provided by the dunes. Mobile dunes and the jagged terrain formed by widespread erosion help to protect animals from excessive exposure to both wind and sun. They can find refuge on the leeward side of any mound, or wallow in the deeper topographical depressions. Dunes provide structure for stabilizing microhabitat temperatures in a region that has so few tree stands and otherwise offers such little protection.

Third (and most intriguing), residents perceive and utilize sand as a
factor of animal husbandry production in its own right. Goat herders of
northern China have long engaged in a practice known locally as *can
shazi* (adding sand). When they shear goats in late April to collect the fine
short hairs from which cashmere is produced, local herders universally
grind sand into the shocks of hair. They do this to inflate the weight of
their produce and thereby increase their market earnings. They have developed a method that efficiently binds sand to hair so that even after the
shocks dry out, the sand will not separate. In fact, it remains remarkably
inconspicuous. The buyers are not oblivious to the practice—they assume
a certain percentage of adulteration and factor that into their purchasing
price. The supplier's challenge, therefore, is to compact more sand into
each shock than the buyer expects. Herders who are most skilled at this
practice typically buy up shocks from their neighbors and restuff them
with sand before selling again to profit from the narrow margins of their
"value added." The sand that is most prized for this purpose is the fine
white sand with no humus. Thus, even when sand is considered "dead,"
it is not altogether barren in this economy.

When it comes to adulteration, the various marketing structures that
have operated in pastoral regions for hundreds of years have all failed to
provide consistent negative feedback linkages between consumer and
producer. One humorous herder made an explicit reference to sand as a
commodity for trade in the international market. He first asked me
whether the United States had all the desert it desired. He then poured
another handful of sand into his sack of cashmere as he joked, "here, I'll
send some more." Without direct feedback linkages, sand and the overall
dusty environment have distinct economic value. As Longworth and
Williamson (1993: 312) noted:

If farmers were penalized for incorporating dust in the wool, they would have an
incentive to avoid dusty environments. At present, the marketing arrangements
encourage wool-growing households to consider dust as one of the "inputs" in
the production of wool. Indeed, one could envisage a "dust production function"
from which an optimal input of dust could be determined and, by implication, an
optimal amount of pasture destruction to provide the dust needed to "produce"
the wool.

Actually, the potential profit margin derived from a "dust production
function" is greatest among communities that specialize in goat husbandry and cashmere production, rather than among sheep herders who
produce the heavier commodity of wool. At 1994 spring prices, a kilo-

gram of sand embedded within a sack of cashmere would earn 200 yuan from outside traders, whereas the same kilogram buried in wool would only earn 10 yuan. This helps explain why residents increasingly specialize in goat husbandry as the range increasingly erodes.

Based on an educated estimate that each kilogram of marketed cashmere contains between twenty and twenty-five grams of sand, and each kilogram of wool contains perhaps five grams, a rough extrapolation based on the goat and sheep population in the village would indicate that Wulanaodu herders must have sold about 1,160 kilograms of sand in 1993.[5] If that ratio of sand to fleece could be consistently applied to all of Inner Mongolia, approximately 3,490 to 3,594 tons of sand would have been marketed in the single year of 1990.[6]

Despite the speculative nature of these estimates, it is clear that up to some hypothetical saturation point, sand is a resource and a desirable feature of the regional landscape. Further, residents could not actually market so much sand if it were not ubiquitous. In fact, a good part of the saturation occurs entirely without human effort. Residents undoubtedly have more sand surface than they need or want, but pockets of "desertified" areas do make positive contributions to local production, and may in fact be necessary to sustain the viability of the economic system as a whole.[7]

Fourth, I should point out that dune sand has more utilitarian value even within the home than is generally assumed. One obvious example is that sand soil functions as an important building material. It provides the core ingredient of mud-brick for house construction, but serves in a great variety of miscellaneous purposes as well (such as wall plaster). Mongol herders also value sand as a material for purification. Humphrey et al. (1993: 53) report that contaminated water may be approved for consumption through the exercise of religious rituals that involve sprinkling it with sand. Mongol women in Nasihan use sand in the cradle and in the swaddling clothes of newborn infants.

## Symbolic Utility of Sand

Mongol herders also relate to their environment through metaphorical symbolism that helps ground their sense of collective identity. In particular, community members sometimes romanticize the sandy environment out of a nostalgia for a fading traditional lifestyle. They wistfully identify with the sand in terms of its persistent mobility. As enclosures proliferate, some residents actually express poetic envy for the "freedom" (*ziyou*) of the swirling sand. One of the village elites with a solid hold on privately fenced pastureland explicitly told me that Mongols had great "respect" for the mobility of sand. While residents grumble about the increasing obstacles to their own free movement, the sand continues to move at will. Along the same nostalgic lines, many people worry aloud about the day when all the land is enclosed and animals have nowhere to roam. One man, who seemed to be complaining and reminiscing at the same time, observed: "the more we settle, the more dust flies."

Such mental associations with the land should not seem strange or exotic—desert sand has certainly conjured powerful symbolic imagery in Western cultures as well. One need only think of the countless sagebrush poets of the American West to appreciate the evocative power of sand-covered terrain. One of the most celebrated apologists for the arid range was Edward Abbey (1984: 77–80):

Sand is nature in the nude, simple, severe, bare. . . . With forms and volumes and masses inconstant as wind but always shapely; dunes like nude bodies, dunes like standing waves, dunes like arcs and sickles, scythe blades and waning moons; virgin dunes untracked by machines, untouched by human feet; dunes firm and solid after rain, ribbed with ripple marks from the wind . . . sand and beauty, sand and death, sand and renewal. . . . Sand dunes make for a simple but varied beauty, as their shade of color changes from hour to hour—bright golden in morning and afternoon, a pallid tan beneath the noon sun, platinum by moonlight, blue sheened under snow, metallic silver when rimmed with hoarfrost, glowing like heated iron at sunrise and sunset, lavender by twilight.

Other scholars have documented the symbolic power of sand in twentieth-century American communities. For example, Engel (1983) explored the way a town in the state of Indiana manipulated the national interpretation of local dunes in order to preserve them and their own sense of collective identity. Through the power of metaphor, sand came to personify the most profound of local ideologies—democracy itself:

Dunes are intensely human, they organize, mobilize, get together, decide to move, and overnight they are gone. . . . Sand—an infinite number of particles of sand—the most elemental and primordial of democracies! One picks up thousands of them in a single handful. Yet each grain is unique, as is each configuration they make together. . . . They are endless variations on the theme of unity in variety. (118–122)

Consistent with this positive symbolic identification, there derived an intense aesthetic appeal: "One need not be a scientist to see the beauty of the Dunes, to sense the drama that is taking place, to feel the surge of some creative force beyond his ken. The dunes are more than sun and wind and sand; they are symbolic of the struggle which all living things endure in order to fulfill their destiny" (119). The poetry of Carl Sandburg and others inspired even grander metaphorical perceptions that came to influence the national debate over whether to preserve the Indiana Dunes. Sandburg saw in the moving mountains of sand the poetic unification of the most primordial opposites—time and eternity. As Engel summarized, "Time is in the process of being written into the landscape while eternity is forever written there. Infinite movement—infinite repose. . . . The Dunes are a changeless yet ever changing plain . . . a creation at once perfect and incomplete" (118–119). Just as sand can be a particularly potent symbol for modern Americans, it is not likely to be less so in out-of-the-way "anthropological places."

Nasihan residents have other significant symbolic associations with dune sand. In general terms, they consider it to be a constituent element of their ethnic identity and way of life. For example, during household interviews, I asked people to explain why Mongols tend to live in areas characterized by sand. No respondent ever challenged the premise of the question. Most of them answered by reference to human and animal population growth. Others made veiled references to the biases of the national political economy, Han colonialism, and historical experiences of ethnic exploitation. I was surprised, however, when a few respondents specifically asserted that their ancestors had pursued the sand as a preferred environment. From their point of view, residents have not suddenly found themselves living in a desert-steppe environment today merely because they overgrazed the range, but because the utility of the diverse landscape had long ago beckoned herdsmen to settle in the area. Conditions on the range have since deteriorated, they concede, but the land was selected through a deliberate historical process.

At least some herders use sand to think about their collective identity in another sense. They perceive a symbolic connection between the overgrazed landscapes of home and their own social marginality within the national political economy. Just as the land is perceived to possess great diversity and potential, some people tend to see themselves as an untapped national resource that is neglected but worthy of investment. This connection suddenly occurred to me one day during an interview with a surprisingly outspoken young man. I asked him to estimate the recovery time of the "worst" landscape photo and he responded by asking me how long the Communists had been in power. He then made the point that the land could recover in just a few years, "if it only had a little help." But since there were no meaningful institutional supports to promote soil conservation in marginal territories, the land has still not recovered its productivity, even after four decades.

This sentiment was expressed in other ways as well. I heard the term "colonization" (*kaituo zhimindi*) used frequently in reference to central government policies that promoted Han in-migration and the exploitation of regional resources. Residents also often complain about cattle from nearby Han farming villages that trespass onto their rangeland. They would not mind if the grazing opportunities were reciprocated, but since the farmers have no pastures to share with Nasihan herds, the practice is deeply resented. One resident became so agitated talking about it that he suggested the Han were something of an epidemic whose coming and going was all but impossible to control, "as you yourself must know from the China Town in your own New York City." Thus, direct connections do exist in the mind of many residents between local land degradation processes and the larger social realities of the Chinese state. Dune sand is by no means always regarded in positive terms, but whether residents perceive it negatively or positively, it consistently reminds them of their own ethnic differentiation.

Lattimore (1941) and more contemporary sources refer to the popular legend of "singing sands" among Mongol herding communities in some regions of Inner Mongolia. Under favorable atmospheric conditions, dune sand emits a loud and distinctive humming sound whenever pedestrians set off a ripple of disturbance. This natural phenomenon provides another vehicle by which sand functions to preserve memories of traditional Mongol culture and thereby sustain a collective identity. Some communities explain the eerie sound by reference to an ancient lamasery buried under the dunes. The spirits blow their trumpets and ring their bells every time somebody marches over the hill, "to let the people know that they are there" (Serruys 1980: 102). In a somewhat analogous fashion, the presence of dune sand serves contemporary Mongol herders as a tangible marker of ethnic and lifestyle difference that continues to remind the surrounding Han Chinese of their enduring presence.

To summarize, in Nasihan the presence of dune sand on the landscape seems to signify for both Han Chinese and Mongol pastoralists the true essence of their historical relationship, though the interpretation of that essence differs for each group. For the dominant Chinese state, the desert continues to signify the primitive nature of the steppeland and its ethnic inhabitants. For Mongol herders, the desert communicates a more complicated and multivocal message. At the most positive extreme, the desert appears nurturing. Local herders recognize dune sand (at both a regional and landscape scale) as the enabling ecological parameter that guarantees the survival of extensive land use systems and sustains the viability of traditional pastoral lifestyles. In this sense, residents legitimately consider sand to be the constituent element of a preferred home environment and familiar way of life. At the most negative extreme, the encroaching desert signifies the bitter legacy of Mongol subjugation under an exploitative political economy and a hostile Chinese state. Sometimes the people seem to appreciate their physical and social separation, sometimes they seem to resent it.

## Fences and Placemaking

Like grass and sand, fence-wire is another landscape feature highly charged with social meaning. Individuals, households, officials, and entire villages all pay close attention to the purchase, placement, erection, desecration, and restoration of private fence-wire in the region.

Nasihan residents do not generally look upon new fences in neutral terms. Nor do they usually reflect upon new enclosures in the benign context of "dune fixation." Rather, they tend to reflect in very personal terms about how it will affect their immediate social relations. For example, they may perceive a neighbor's fence as a challenge of entitlement ("Hey, I was going to fence that land!"), as a threat ("You are trying to cut me off!"), as ostentatious display ("When did he get so much money?"), or simply as a nuisance ("Now I must go around this damn field!"). But residents also reflect upon new enclosures in more abstract terms that relate to their collective identity. For example, they look at a neighbor's fence and perceive issues such as community threat ("That expansion just brings us all one step closer to the extinction of open-range grazing"), or government corruption ("How did he get permission to claim that land?"), or administrative failure ("It's everyone out for themselves"), or local nepotism ("He got a loan but I did not"), or exploitative state relations ("We have no production support services"). I heard residents express all of these perspectives at one time or another.

Residents also correctly understand the fence as a tool of external control. They know that enclosures impose restrictions on themselves as well as their livestock. Grassland scientists, for example, increasingly assume the authority to set parameters of household production. The following published quote offers a good example:

Scientific research can determine the optimal number and type of animals to graze in a particular area, during a particular season and period of time. . . . A rational rangeland management system, coordinated by central and local government agencies and backed by regulations and policies, is required to make sure that each rangeland area is properly utilized. (Zhang 1992: 49)

Furthermore, the prosperity of an entire region can rise or fall with scientific pronouncements over production potential that influences the relative size of government investments. For example, some outspoken scientists now favor investments in southern and western grasslands over those of arid Inner Mongolia (NRC 1992: 140). The influence to control large populations and to shape regional prosperity is an obvious manifestation of the increasing political power of grassland science in China. It is no wonder then that the majority of residents feel vulnerable and even resent some of the transformations set in motion by government policy.

## RESENTMENT AND RESISTANCE

The unpopularity of grassland policies only increases local opposition to Han scientists and their practice of grassland science in Nasihan. First, there is much anger directed toward the itinerant scientists as individuals. Locals interpret the research station primarily as a boondoggle. They believe that little has been accomplished after twenty-five years of intervention and great expenditures of capital. In the words of one outspoken resident, "only about five years have been beneficial for the community, the other twenty years have been a waste." Many people claim they could personally do a better job if given just a fraction of the funds. Some even charge the scientists with making bad conditions worse. For example, they mock them as idle tourists and "guests" (*keren*) who hoard the best lands for their own garden and nurseries, living quarters, and experimental fields: "If they came to control the desert, should not our guests live out amongst the moving dunes rather than in the lap of luxury?"

Others criticize the Han scientists not only for mismanagement but also for ethical improprieties. They charge them with various forms of economic exploitation and opportunism. For example, the research station often hires local workers (including children) at a meager wage to provide much of the physical labor required to maintain the grounds, to carry equipment, and even to collect and record field data. The scientists also participate in the local economy by purchasing sheep, which they entrust to local herders for years at a time. This leads to unpleasant confrontations (which I witnessed) when the absentee owners suddenly appear and demand full accounting for the herd and its natural increase. It was reported to me that some members of the scientific community had even engaged in the trade of local livestock, buying low in the village while selling high in the cities for personal profit.[8] For these and other criticisms, most of the residents would like to see the station closed down.

There is also animosity directed toward grassland science in general, which takes both passive and active forms. In polite conversation, local residents all repeat formulaic praise for scientific methods of production. Yet the vast majority of them quickly admit that they are not scientific practitioners. They verbally praise the ideal, but feel content to persist in their "backward" traditional ways. When challenged to explain this contradiction, residents eagerly defer such conversation to their local "model" citizens. Local herders are willing to praise these token households for production practices that they personally have no intention of adopting, apparently for the insulating cover it provides them from the scientists.

More active forms of opposition to grassland science can also be observed. For example, herders do not usually implement the new rangeland policies as intended. They do not follow a wide range of enclosure specifications, including the primary injunction to keep livestock contained within household fences. They generally do not subfence their land or practice coordinated rotational grazing. Some individuals still cut down trees for timber, uproot sod for housing, and cultivate gardens in excess of 30 mu—all illegal activities under national grassland regulations. They prefer indigenous breeds of sheep to exotic species, and despite explicit grassland prohibitions, they refuse to abandon or even limit their goat husbandry.

Herders also denigrate grassland science on grounds of ineffectiveness. They complain that farmers enjoy the benefits of improved varieties of rice seedlings while they must do without drought-resistant fodder crops. They ridicule aerial seeding as impractical and a waste of resources. They consider artificial insemination to produce weak and sickly animals. They perceive no meaningful government investment in the pastoral livestock sector. Indeed, they point to the hypocrisy of government officials who sound the alarm of degradation yet continue to resist substantial investments in the land and its people. In short, they not only reject the Chinese national discourse that would scapegoat Mongol herders, they explicitly blame Beijing for causing (through colonialism) and perpetuating (through neglect) the land degradation that jeopardizes their live-lihood.

There is also much vandalism explicitly directed at research station property and science equipment. In particular, fence-wire is frequently stolen and weather recording instruments are damaged. Even more threat-ening to the practice of local science, there is much deliberate human and livestock incursion upon the enclosed fields where scientists conduct their research experiments. Angry that so much land has been taken out of pro-duction for so long to serve as an experimental control, herders frequently graze and mow vegetation within these boundaries. In so many ways, lo-cal residents contest the rigid walls of separation that grassland science has erected to reproduce fundamental social hierarchies in China.[9] The Han scientists certainly know that they work in an antagonistic relation-ship with local residents, but they perceive the various expressions of re-sistance as further evidence of the "backward" pastoral culture that keeps the region degraded and impoverished. They speak in terms of native res-idents being unable to cope and cooperate with the scientific mission. Every negative encounter with the local population simply reconfirms for them the official discourse that scapegoats local land users.

SYMBOLIC SABOTAGE

Beyond the routine acts of vandalism directed at the scientists and the frequent feuds between neighbors that new fences have motivated, I also witnessed how violations of private enclosures sometimes soared beyond the mundane to become highly symbolic acts of significance that cap-tured the attention of the entire community. The best example is a con-flict that involved the Han scientists at the research station and a local fa-vorite son named Muergen.

In recent years, Muergen has been an important affiliate and channel of ingress into local affairs for the research staff. He is about forty years old and lives just across a dirt road from the front entrance to the re-search station, in a house that was originally built by and later sold to him by the senior director of the station. Like the community at large, Muergen and the scientists have a long-standing but ambivalent relation-ship. They often cooperate for mutual benefit, yet each deeply distrusts the other, not only because of class and ethnic differences, but also be-cause of perceived conflicts of interest over the use of local resources. Fluctuating relations with Muergen tend to serve the staff as a measure for the state of their relations with the whole village.

For many years now, Muergen has accepted token wages (2.8 yuan per day) to protect the station assets from vandalism. No resident would dare violate research property with Muergen as caretaker because of his clout in local politics. He is a highly influential member of the Wang family, older brother to the village Party secretary, cousin to the village chief, nephew to the chief of Nasihan township, cousin to the township police chief, and himself an elected village representative. In addition, he is highly respected throughout the township as an intelligent and hard worker, and he is widely admired for being witty and outspoken (his name means "clever" in Mongolian). His work contract with the scientists might alternatively be construed as a bribe, since it requires no specific labor (they also employ a full-time Han groundskeeper), and since he might otherwise consider it in his self-interest to oppose the station's work. In addition to the wages he earns, Muergen also enjoys the exclusive privilege of cutting hay on several hectares of experimental meadows. Despite these benefits, Muergen is neither a lackey of the scientists, nor an apologist for their activities in the village. On the contrary, he is occasionally their sharpest critic over specific management issues, precisely because his influence carries weight with them just as it does with the villagers. He thus enjoys a unique role as mediator that gains him social prestige in both camps.

It created quite a sensation, therefore, when it was revealed that Muergen himself took a sledgehammer and smashed down the rear gate of the research station wall one rainy night in order to lead a herd of fifteen cows to feed on the luxuriant garden enclosed within the compound. The following day, the junior director of the station surveyed the damage and easily tracked the hoof prints back to the property of Muergen. He then called for a police investigation, trumping Muergen's local clout by appealing to banner-level officials. They also followed the evidence back to Muergen, whereupon he admitted to breaking down the gate and immediately paid a fine of 500 yuan. His contract with the station was terminated.

The interesting question is: Why did he do it? Muergen derived no economic advantage from this deliberate sabotage—his cows were not without supplies of hay, and they only consumed a few rows of corn and sorghum anyway. He did, however, gain much symbolic capital in the community from an action that most people interpreted as noble ideological defiance. The point, obviously, was not to feed the cows, but to desecrate the compound, the garden in particular. The garden had become a sore point in recent months when the junior station director decided to improve cash flow by contracting the large irrigated plots out to Han service personnel at the station (the cook and the groundskeeper). When word leaked into the village, residents felt insulted that the land had not been awarded to local households instead. The decision was perceived as just another example of Han chauvinism and colonial conquest.

Although some residents did not support Muergen's action, everyone understood it as a brave assertion of local identity—Mongol over Han, herder over scientist, worker over intellectual, resident over colonizer. His action was translated into words for me by one supportive neighbor: "The station exploits local residents by removing good land from local production. It is time for the Han to leave and return all the land to its rightful occupiers." Muergen's attack surely had multiple dimensions of motivation, but whatever else it may have communicated, it was in fact an assault literally against the erected wall of separation. I suspect that Muergen was emboldened to commit the symbolic deed precisely because it was most meaningful coming from him. Though prosecuted, Muergen was applauded by many residents for at least temporarily reversing the established order and restoring enclosed land to pasture. For a community of Mongol herders, no symbol could better make that statement of opposition than a desecrated compound wall.

### SYMBOLIC APPROPRIATION

Muergen's story confirms that spatial barriers and their desecration evoke powerful symbolic meanings for the community. I do not mean to imply that those symbols always and only evoke negative or hostile feelings. I would suggest that on some level, conscious or unconscious, Nasihan residents have discovered a way to incorporate fences more constructively into their own symbolic life-world. Indeed, they have little choice. In essence, they apply indigenous values to the technological materials that have been thrust upon them by the world outside, and thereby they creatively finesse the inevitable transition between a traditional mobile lifestyle based upon extensive grazing and a future sedentary lifestyle based upon intensive land use. They achieve this transition in symbolic terms by transforming the fence's essential signification from an image of sedentarization to one of mobility. In Nasihan, fences do not function to keep animals inside an enclosure, they function to keep animals out. They do not, contrary to policy, attach livestock to space. Rather, they guarantee their rotation through space.

Cohen (1985: 28) has remarked upon the "resourcefulness with which people use symbols to reassert community and its boundaries when the processes and consequences of change threaten its integrity." Using a metaphor that herders would find comforting in its familiarity, he argued that social change is often marked by a regurgitative process that amounts to a veiled refusal to swallow:

Like a cow chewing its cud, a community tends to adopt new social structures that originate from outside its boundaries only by slowly transforming them in the process of importation, fundamentally reconstituting them with indigenous meaning. . . . In this way structures imported across the boundary provide new media for the expression of native values. . . . People can turn these alien structural influences to the service of their indigenous symbolic systems and thereby symbolically reinforce their customary boundaries. (46, 75)

From this point of view, the chaotic grazing practices of Nasihan herders do have symbolic, as well as material, purposes. The grazing system and the indigenous use of fencing that enables it are transitional adaptations that permit old grazing patterns to coexist with new ideological and economic realities imposed by a state intent upon modernization. The indigenous fence works as a new symbol of community and ethnic identity because its inherent reference has been changed from one of stability to that of mobility. Thus, the restructured physical environment of Nasihan derives not only from policy decisions made at the national level, and not only from the increasing interference of a globally structured political economy, but also from the intentional agency of local actors.

## Landscape and Hidden History

The critical landscape features of sand, grass, and fence are closely monitored both inside and outside the community. But these symbols are interpreted locally within the context of existing built structures already on the landscape, most of which bring to mind a long history of troubled relations with the Chinese state. In such company, the grass-sand-fence constellation can hardly elicit much enthusiasm for the future.

Here, in review, are some of the more evocative images emanating from the contemporary local landscape, and the hidden histories to which they refer, that provide running political commentary for native inhabitants:

—the memorial stones that mark the graves of Communist soldiers
  murdered for pillaging local livestock, which reference the long War

of Liberation and the ambivalence of the local population toward
occupying armies

—the Hongshan Water Reservoir, which references both the Great
Leap Forward and the regional deforestation that resulted

—the dysfunctional irrigation canals that disguised an underground
garrison, which reference Cold War tension with the Soviet Union

—the multiple military structures in the area, which reference both the
fears of the Cold War and the marginality of the region

—the colossal watering well that commune production teams were too
fearful to protest or amend, which references the waste and terror of
the Cultural Revolution

—the scattered stones on hilltops that formed religious shrines until
Red Guards demolished them, which references the violence and the
suppression of ethnic identity during the Cultural Revolution

—the high insulating wall of the research station, which references
Han colonialism and ethnic elitism

—the proliferation of fence-wire and the concentration of grassland
resources, which reference decollectivization and the end of egalitarian ideology

—the prominent stump of an almond tree that was illegally cut by the
village chief, which references the corruption of local government

—the amputated and frozen bodies of land-poor herders, which reference community poverty, marginality, and ongoing policies of forced
sedentism

No doubt there are many more important symbolic images that reference the past and the present for local residents. This list includes only
those few structures for which I happened to gain some insight into local
impressions. The landscape is obviously rich with imagery that references
both individual and community suffering.

In his seminal study of the English countryside, Raymond Williams
(1973: 144) asserted that social exploitation leaves no visible mark in rural areas, but is "dissolved into a landscape." Tuan (1979: 141) followed
this theme in reflecting upon the great enclosure movement of seven-
teenth-century England. He wrote:

Oppression in the countryside is not egregiously visible and seldom leaves any
lasting imprint. Consider the enclosure movements in England, which resulted in
greater agricultural productivity and in the neat, hedged fields that we have all
come to admire. This was the success story embossed proudly on the land. Behind
it lay numerous tales of deprivation and fear which, but for the literary record,

would have faded from our consciousness because they left their mark largely on the perishable bodies and minds of the people.

Undoubtedly, these authors write from an outsider's perspective. The insider knows very well what the landscape documents. The enclosure movement as experienced in Nasihan leaves a tangible record of suffering on the landscape that residents can easily interpret. The restructured rangeland simply adds another layer of built structure and landscape interpretation that reflects poorly upon local relations with the Chinese state. Considering the violence of the past fifty years, and the great insecurities that modernization drives have inflicted upon ethnic minorities, it is not surprising that Mongol herders take symbolic refuge where they can. It is not surprising that native fences operate in surprising ways, nor that national policies provoke surprising local responses.

# NINE

# Implications and Conclusion

Ecological environments are constructed and transformed by complex and reciprocal interactions between human populations, animal populations, and the physical forces of nature that occur across local, regional, and global scales. At any scale of analysis, these interactions are understood only incompletely, and the great variety of perspectives across many disciplines are all instrumental in the effort to promote human understanding of socially defined environmental problems. Anthropologists can contribute substantially to the effort by situating human decision-making behaviors within specific communities of known individuals to observe how practices of local resource management are both constrained and enabled by powerful social forces that are not necessarily obvious nor material. The attempt to broaden the interpretive framework for understanding human–environment relationships in this way should be welcomed by all.

This book, in addition to the detailed ethnographic information it provides, attempts to refine Western scientific understanding of a number of important social issues ranging from the narrow and empirical questions of grassland policy in China to the more general and theoretical concerns of cultural change, development, and sustainability. Indeed, the real significance of this study extends far beyond the physical stakes of newly erected fence posts in Nasihan to encompass the metaphorical stakes of reconstructing stable environmental relationships throughout the developing world.

## Implications for Rangeland Conservation

The national discourse of "grassland construction" is a political myth. Private enclosures are not saving the grasslands and the government has no justification for blaming current problems of desert expansion on lo-cal producers. The enclosure policy remains in force because it effectively reconstitutes the open range in accordance with the environmental pref-erences and cultural biases of the nationalist Han Chinese. Regimented space replaces open horizons, homogenized fields replace heterogeneous patchwork, resource concentration replaces landscape diversity, and clear boundaries replace the casual mix of grass and sand. The reconstructed range increasingly proves the old Chinese colonial stereotype: land that embodies no labor has no value and is only barren wasteland. Perhaps Chinese officials and scientists continue to obfuscate the local impact of enclosures precisely because they have so much to gain from their lack of clarity. The status quo serves powerful political interests by reproducing a national discourse concerning the frontier that affirms fundamental as-sumptions about the accomplishments of the reform era, the benevolence of the Chinese state, and the superiority of Han civilization.

Despite the overblown public rhetoric of "controlling the desert," Chi-nese experimentation with huge population migrations, technological transfers, and rangeland privatization are not instruments in the con-struction of northern grasslands so much as they are instruments of trans-formation. After more than a century of external interventions that have now culminated in the proliferation of household enclosures, the arid-steppe environment has not been constructed so much as deconstructed and slowly refashioned into an environment that contradicts the ecologi-cal and spatial preferences of indigenous herders. In addition to the social disorientation and general anomie that have ensued, these dramatic changes also disrupt long-standing feedback loops between residents and their production environment.

Visual cues on the landscape that once guided local herders as they made daily and seasonal resource management decisions no longer pro-vide the same information. The indigenous patch matrix has been dis-rupted so thoroughly that residents are less confident about reading the landscape for purposes of production. They are not so clear how to in-terpret the new patterns they confront. Consequently, herding skills have become less relevant and herders have become more apathetic about their ability to control livestock and herd dynamics. More and more, house-holds have no grazing strategy; they simply let the animals wander the range where they will. As local jargon has it, "graze as you please" (*sui-bian fangmu*).

Mearns (1993a: 78) has noted the same general scenario at work among herders in the nation of Mongolia: "The dismantling of the formal institutions of state socialism, and the hardships and political instability of economic transition, have led to conditions of structural chaos in which it is difficult for all producers to anticipate signals in their production environment." In Nasihan, rangeland reconstruction has rendered much indigenous ecological knowledge superfluous. The ability to respond skillfully to subtle environmental cues is becoming a lost art. This fact is the most far-reaching resource management consequence of the new order. When coupled with the optimistic views about land resiliency and sand utility, the erosion of traditional perceptual thresholds could have disastrous consequences for local soil conservation as the transition to a more sedentary, agro-pastoral economy drags slowly onward.

Expressing a similar concern, Longworth and Williamson (1993: 313) have argued that contemporary Mongols of North China have no cultural foundation from which to resist commercial forces that encourage them to accelerate land degradation:

Present day ethnic minority communities in pastoral areas have lost much of their traditional culture. They have been under the influence of Han-dominated governments for at least 45 years and for much longer in many parts of China. The almost godlike status attached to grasslands by their ancestors has gone for ever and in its place is a kind of commercial pragmatism. The danger arising from this move is that if, within a given economic and policy framework, private households are provided with commercial incentives to degrade the rangeland, then social barriers to such actions are likely to be minimal.

Such a danger certainly exists, but there is also room for some optimism. Local cultural expression is highly resilient and could still be channeled in powerful ways to promote an effective—and sorely needed—system of grazing rotation. A revitalized mobility, for example, could both preserve aspects of a cherished way of life and help to sustain the productivity of the threatened grassland ecology. Humphrey and Sneath (1999), for example, have made the explicit argument that extensive mobility actually encourages the most efficient use of rangeland resources, and they report a strong association all across the Mongolian steppe between those areas where native culture is under threat and those where ecological problems are most severe.

It is noteworthy that decollectivization in Mongolia proper has not resulted in rangeland privatization or enclosure. Mearns (1993a: 77) does report some increasing restriction in the total area of pasture to which particular groups enjoy access, but herders still follow an "essentially nomadic lifestyle" (1991: 25). Indeed, Mearns has warned of the dangers of uncontrolled privatization, arguing that the minimal territorial unit required to sustain pastoral livestock production in desert-steppe environments corresponds to boundaries on the order of 3,500 km² (1993b: 85).

But Chinese leaders have embarked on a privatization strategy and will not reverse course in the foreseeable future. The trends I have described here have continued unabated since the conclusion of my fieldwork. Letters from Nasihan and reports from other professionals who have visited the region more recently than I confirm that residents continue to enclose pasture resources, that stratification of wealth increases, that biodiversity decreases, that community discord and alcohol abuse is still prevalent, and that land degradation intensifies. The experience of Nasihan residents would suggest a rather bleak prognosis for the success of grassland enclosure policies. Whatever the original intentions, the way the policies have been implemented at the village level does not promote sustainable pastoral production over the short or medium term. They may yet produce a conservationist effect over a much longer term as the entire range becomes parceled into separate enclosures and some guidance emerges to coordinate rotational grazing. As the situation now stands, however, chaotic grazing practices may substantially redefine the range and the community before any presumed long-term benefits ever have a chance to evolve.

## Implications for Environmental Studies

### RELEVANCE OF PERCEPTION

Over the last decade or so, the scientific model of ecosystem homeostasis has come under increasing attack, and consequently, many conventional ideas about what constitutes order and disorder in the natural world have come under greater scrutiny. In the words of Bocking (1994: 16), "nature is no longer an orderly system in equilibrium; it is instead a patchwork, characterized by pervasive disturbance and instability; constancy has been replaced by change, chaos, and nonequilibrium." In particular, grassland ecosystems, once the very inspiration for steady-state climax theory, have been reinterpreted as prime examples of instability and disturbance (see Loucks et al. 1985; Worster 1990: 10). For the social scientist, these intellectual developments invite a thorough reconsideration of the interplay between patch dynamics and perceptual thresholds within specific social settings. The evidence presented here indicates that some pastoral groups do seem to perceive and think about their production environment in distinctive ways.

Considering the growing concern over land degradation in arid lands, it seems counterintuitive that some pastoral peoples could perceive dune sand as a valued environmental patch. Yet my observations and experiences from Inner Mongolia would support that unusual perspective. Despite the pessimism of the desertification literature, sand does function as both hazard and resource for Nasihan residents. Furthermore, they express positive attitudes toward dune sand within the context of specific patchwork scales. They have more sand surface than they want, especially at the field scale where a national rangeland enclosure policy is gradually destroying the traditional patch matrix of grass and sand with devastating effect for some households. Nonetheless, native tolerance—even preference—for sand is much higher than usually assumed. Considering their distinctive historical and cultural experiences, it should come as no surprise that Mongol herders do not all share the same stark view of their environment as Han grassland scientists, the Chinese state in general, or the broad international network of intellectuals.

Let me be clear: I am not suggesting, nor did any herder ever suggest to me, that soil erosion is not a serious problem in Inner Mongolia that has become increasingly worse in recent decades. My point is that the interpretations, explanations, and solutions are not nearly so clear-cut as the state would have it. Certainly, more discriminating information about dune sand in dryland areas is required to devise the most appropriate strategies for interpreting and dealing with acute land management problems. The challenge to account more honestly for local perceptions would require officials and scholars in China to reconsider much of their standard rhetoric about desertification in the northern grasslands, which is routinely packaged in misleading statements like the following: "Sanddrift, known as the ferocious 'yellow dragon' to local inhabitants, has always stood in the way of the people's livelihood and national construction. . . . The conquest of the yellow dragon has long been the need and desire of those who have lived in its path" (Hsiao 1960: 24). "In the past, the desert was regarded as nothing more than a devil jeopardizing the quality of life of human beings. Now, however, we are devoted to creating an industry [of fruit, grain, and fisheries] based on the sand" (Jiang Wandi 1994: 20). Such simplistic and politically motivated caricatures of an unfamiliar pastoral environment and lifestyle do not promote institutional understanding of complicated land management issues. Native attitudes and perceptual thresholds may in fact turn out to be critical assets in the search for long-term sustainable resource management strategies. Outsiders would do better to consider them respectfully in the context of diversity and opportunity, rather than dismiss them outright as residual cultural baggage.

RELEVANCE OF POLITICAL FEAR

The high degree to which Chinese agrarian policies have been politicized in the past contributes substantially to unsustainable local land use even today. Village cadres, sensitive to political risk, hesitate to align themselves closely with national campaigns or prevailing land management doctrine. For example, the post-reform political climate that denounces former Maoist policies reminds local administrators and household managers of the risks of betting too heavily on the ideological descendants of Deng Xiaoping, market reforms, or currently fashionable land use policies. The historical legacies of political risk therefore continue to interfere with the possibilities for effective state control over socially undesirable land use practices. The reverse is also true—the legacies interfere with the possibilities for effective social control over undesirable state land use decisions. Memories of violence and personal loss interfere with the willingness of local producers to provide necessary feedback to policy makers. They also interfere with the motivation of herders to invest heavily in easily confiscated land resources.

There is, however, another sense in which the enclosure movement on Chinese grasslands ultimately facilitates the expansion of state surveillance and social control far beyond the narrow concerns of land degradation. Enclosure is neither a haphazard nor an innocuous change in land use policy, but a critical acceleration in the greater modernist project to extend the reach of governmental authority over a subject population. The goal is to regiment local processes of production and to reorganize the daily activities of rural producers in the service of a more efficient flow of global capital. Foucault (1979: 198) laid the foundation for this argument when he carefully explored the history of relations between political power and the organization of space. He conceived of modern landscape as a political technology dedicated to creating passive citizens and docile bodies that could be more easily subjected, transformed, and improved for more efficient use by state institutions. Foucault observed that in the long transition from traditional institutions to modern bureaucracies, state power increasingly exercised greater authority over individuals by regimenting the organization of geography and fixing individuals within bracketed, observable, and therefore more controllable spaces—like schools, hospitals, prisons, military barracks, factories, and just about any locus of production.

Foucault and others generally date the origin of this pervasive manner of modern surveillance to the mid-1800s, when expanding state powers in Western Europe no longer rested content simply to take a share of what was produced and exchanged through taxes, but began to assume a more total presence by subtly infiltrating into the processes of production itself (Foucault 1980: 78–108). The first step in that process, in the language of contemporary geographers, is the replacement of a "space of places" by a "space of flows"—the local must yield to the global, the particular to the universal. The new order of production ensures that local behaviors increasingly fall under the influence of activities and decisions occurring in distant centers of power. Ironically, the more state administrators manage to impose a uniform and homogenous spatial order, the more they can dismantle the obvious and brute manifestations of social control. The new order basically runs on automatic pilot once a community has been harnessed for service and subdued by their fears.

RELEVANCE OF CULTURE

Landscape transformations in Inner Mongolia cannot be fully explained or understood through paradigms that consider only the natural, long-term evolutionary dynamics of grassland ecosystems. Anthropogenic pressures have been too intense and too disruptive for such a reductionistic approach. Nor do political-economic considerations explain all that is relevant to the observed patterns of land use. The historical-materialist perspective certainly advances our understanding by showing the relationship between the exploitation of local land managers and the exploitation of land in the context of intrusive forces that seek to link village residents more directly with distant centers of power. Yet this popular framework tends to ignore the manifest social meaningfulness of landscape symbolism, group identity, and cultural creativity. Mongol herders do not simply surrender their familiar life-world in the process of economic development. On the contrary, they actively recreate territorial coherence on ecological and spatial terms that are acceptable to them. Indeed, distinctive group perceptions of space and place may in fact be constituent elements of the transnational "culture of nationalism" that has proven so resilient throughout the developing world in the late twentieth century.

To say the least, ideologically informed perceptions, fears, and motivations are highly relevant in shaping and contesting the national policies and the local practices that have historically structured—and continue to restructure—the ecological environment of Nasihan. I have provided evidence that local landscape is a social construct that bears the tangible marks of power and conflict. It is at once a historical product, a social determinant, and a continuous medium of expression between and among various groups. In short, Nasihan residents (and presumably other herding populations) have their own cultural perspectives that do not accord so well with a national discourse that obsessively associates economic and moral deficiencies with arid rangeland. Nor do they accord so well with the international scientific community that seeks to privatize common property resources and impose a "rationalized" production regime all over the world. A fair accounting of local circumstances will force scholars seriously to qualify, if not outright contradict, much of the prevailing rhetoric and conventional wisdom in China about the situation on the grasslands.

Fuller consideration of the cultural dimensions of environmental change will require supplanting simplistic explanatory models from the past, which tend to emphasize singular causes for resource management problems such as population growth, communal land tenure, or economic exploitation, with more nuanced processual models that actually explore the messy circumstances of social reality—including the agonizing tensions between economic growth and stratification, accommodation and resistance, development and subjugation. A more comprehensive accounting for cultural and social process will also challenge rigid disciplinary boundaries that continue to separate the physical and social sciences, that help to privilege nonlocal representations of nature through the exercise of social power, and that work to conceal the subjective dimensions of scientific knowledge construction. Only by working beyond the familiar nature/culture dichotomy—that most substantial of intellectual walls known to scientific enterprise—can we pursue a dialogue among competing knowledge systems that must occur in order to improve our understanding of environmental transformation.

Ultimately it matters to the construction of scientific knowledge that problems of interdisciplinary collaboration at the international scale lead to Western natural scientists working exclusively with Chinese natural scientists in a minority field setting. Unhealthy dependencies are easily created that distort data collection and close off opportunities to engage an alternative knowledge base. It matters if Western scientists tour research enclosures to monitor the success of dune fixation without any knowledge that such enclosures are fiercely contested (and vandalized) by the residential community. It matters if they concern themselves with land degradation and resource development but never learn that Mongol herders have their own conflicting definitions of these concepts. It matters if they visit local households to evaluate community needs without knowing those individuals to be "model citizens" who are engaged in well-rehearsed (and duplicitous) role playing. It matters if they utilize ecological and weather data collected and recorded by local workers who feel exploited by and hostile to their absentee urban employers. It matters because the field data may be wrong, or incomplete, or significantly context dependent.

The arguments presented here should not be construed as an attempt to reject scientific practice. On the contrary, they underscore the need for carefully chosen research strategies and the importance of persuasive evidence. The arguments, however, are intended to help indict the pernicious and all too widespread practice of scientism—the arrogant assumption that scientific knowledge necessarily exists as objective fact, independent of social and institutional context. Ethnographic fieldwork helps to demonstrate that competing social values and institutions are often essential facts to consider in resource management problems.

## Implications for Development

It is now a matter of conventional wisdom that the Western scientific understanding of nature as an interdependent global ecosystem has established the intellectual foundation and political rationale for coordinating environmental conservation measures and streamlining public discourse at an international scale. I think it justified to use the term "environmental regime" to identify what appears to be a powerful integration of cultural assumptions, scientific understandings, and administrative institutions that mutually influence each other on a global scale to reconstruct local space and environmental conditions, often contrary to the immediate interests of the majority of native residents. This usage is consistent with other recent scholarship which speaks of influential "epistemic communities" (Haas 1989; 1992), and "intergovernmental environmental domains" (Frank 1997), and "world associational arenas" (Meyer et al. 1997). The emergence of a global environmental regime is usually understood as a positive development because a shared scientific conception of nature is deemed necessary to control the grave and multiple environmental issues that now cross national boundaries. Yet there are also some important costs associated with such a regime that do not receive sufficient attention.

One significant cost is the fact that an international language of re-source management necessarily eclipses the parochial perspectives and concerns of local residents, who may in fact have a more nuanced appreciation of sustainable human–environment relations in a particular locale. In Inner Mongolia, Chinese and international officials have imposed their own values and economic agendas on minority populations who do not share their view of nature or vision for the future. The majority of residents do not endorse current land use policies because they legitimately see them as detrimental to their own survival. Yet environmental regimes are programmed to focus primarily on issues of "compliance" with prescribed rules and regulations, rather than on the sometimes very sensible reasons for and social context of noncompliance. Prospects for overcoming serious resource management problems are thus hampered rather than promoted.

A related cost is the fact that incorporation within the global political economy tends to reduce the biodiversity and stability of local production environments. The price of development is conformity, which involves a process of self-denial that many groups find intolerable. As Shiva (1995) has pointed out, the global system actually needs to maintain the marginality of some social groups in order to preserve the possibility for biodiversity at a global scale, yet the continual penetration of international market forces and "rational" resource management principles into the developing world threatens to eradicate the very diversity that draws their attention. The recent experiences of Nasihan herders support the contrarian view that the restoration and sustainability of biodiversity in the modern world will require decreasing, rather than increasing, linkages with the global system.

Another significant but underestimated cost of global oversight is the unexpected damage that may occur during the "transition" phase of new policy prescriptions. For example, land reforms take time to implement, but the social chaos that occurs during the transitional period may well trigger a destruction of local resources that has greater environmental impact than the amelioration that is presumed to follow. In Nasihan, range-land privatization has exacerbated land degradation processes over a time frame of sufficient duration to merit serious concern about the future prospects of the land. The long transition time also carries negative consequences for the pursuit of social justice and production security.

Ultimately, this research raises significant questions about the nature of "discourse" and the relationship between empirical evidence and ideology. Are public discourses mere fictions produced by the political interests of different social groups, or do they reflect more tangible realities of experience that can be neutrally evaluated through the exercise of reason? I am inclined toward the latter view and optimistically believe that there does exist a willingness on the part of different knowledge communities to listen to each other. Significant progress in resolving real-world problems can be achieved simply by exposing the competing perceptions and knowledge bases that otherwise remain concealed. The practice of scapegoating local populations through the guise of an objective and rational science does not advance the cause of either science or development.

## Concluding Thoughts

The flow of global capital has been restructuring local experiences of space and physical environment throughout the world for centuries. The point of this book is not merely that Nasihan represents the latest chapter in a rather old and predictable story of social transformation. Rather, it is that each modernization experience can be highly significant and informative because of the resilient agency of local culture. Since there are no passive agents in the process of transformation, local residents actively and creatively engage global disciplinary influences on their own cultural terms. There is no reason to expect the experience of modernization to be uniform either in process or in outcome. Rather, forces that are global in structure produce unique outcomes in parochial settings precisely because they unfold under varying circumstances of history and ideology. Nasihan township deserves our attention because its particular circumstances serve to illuminate with unusual clarity many of the significant subprocesses typically involved in the experience of modernization: stratification, commercialization, urbanization, alienation, globalization, environmental degradation, health risk redistribution, and time-space reorientation.

The turmoil under way in Nasihan should touch everyone personally because we must all come to terms with the perpetual changes that condition our own modern living. Toynbee (1967: 28) once discussed the lost sense of place among modern inhabitants of sprawling cities in terms of a deep existential crisis: "Megalopolis is going to swallow up human beings of all cultures, religions, and races; for all of us, the problem of having to live in Megalopolis will have to be solved in spiritual terms." To avoid placelessness and inauthentic living, local actors are increasingly compelled to redefine themselves anew within a pervasive world system. How and why they do this concerns us all.

Many of the transformations that modernization brings may well be inevitable. Yet the question of how to make the process less painful and traumatic for the people involved is a significant development issue worthy of greater deliberation. Only when the views of rural producers themselves are taken seriously into account can we begin to expect their enthusiastic and successful participation. The story of Nasihan shows that local communities and individuals are viable players in the process of development whom governments cannot afford to disregard.

# Notes

### Chapter 1: A Land and People in the Way

1. Since 1979, China has adopted an ambitious family planning program known as the "one-child policy." In a dramatic attempt to control population growth and facilitate economic development, the reform government set up a large bureaucratic apparatus to oversee and enforce unpopular birth restrictions. Yet it is inaccurate to view the one-child policy as a set of stable and unchanging proscriptions. There have been periods when the laws were applied rather harshly, as well as moments of relative relaxation. Also, since the policy is implemented by regional and local authorities, it has been applied differently across the nation. Urban Han Chinese couples are generally permitted to produce only one offspring, no matter the sex of the child. Rural Han Chinese are permitted to bear a second child if the first is a girl or physically impaired. Urban couples who belong to a recognized national minority are permitted two children, regardless of sex, and minority couples in rural areas are permitted up to three children if the first two are girls and if the parents adhere to a birth-spacing of several years.

In general, rural minority populations have been virtually exempt from the restrictions until the late 1980s, when harsh penalties began to be applied. There are a variety of circumstances under which exceptions to the policy may be tolerated, but households stand to gain many benefits through compliance and face many penalties for noncompliance. The policy allows for imposition of a fine for the household, forced sterilization for the mother, disqualification from a range of public services (such as education) for the child, possible demotion or other discipline for local bureaucratic officials, and sometimes even financial penalties for an entire community. (For a critical analysis of the policy, see Aird 1990; for a more ethnographic treatment, see Croll, Davin, and Kane 1985.)

2. Walter (1988: 23) introduced this term to describe a "radical shift of topistic structure, a fundamental change in the form of dwelling together . . . [that] conceals, interrupts, or breaks the old forms, causing new structures by patterns of exclusion, enclosure, and dissociation." Topomorphic revolutions transform traditional common living spaces into more complicated geographic systems of segregated zones.

3. Of course, stocking ratios and other such measures are indeed essential tools in the analysis of rangeland management, but they are incomplete without

analysis of the social context of land use or the decision-making apparatus of local resource consumers.

4. This is not terribly surprising, since the formal study of sociology was discontinued within Chinese research academies from 1952 until 1979, and only one university since 1980—Beijing University—has organized a program of sociological research in the grasslands (Ma 1992: 122–123).

5. It is reasonable to wonder why scientists from the city of Shenyang in neighboring Liaoning province hold jurisdiction over rangeland in the Inner Mongolia Autonomous Region. As best as I can determine, the arrangement seems to follow from a series of administrative changes initiated by Mao in the late 1960s under the pretext of national defense. After Sino-Soviet border clashes in 1969, the Shenyang Military Region of the People's Liberation Army assumed responsibility for the defense of China's entire northeastern region. Because of favorable transportation linkages, Eastern Inner Mongolia was placed under Shenyang's sphere of influence. The military jurisdiction then carried over into many other administrative responsibilities, including environmental protection.

6. To my knowledge, only one scientist was a member of an ethnic minority group (not Mongolian) within China. Since the Shenyang Institute is located outside of Inner Mongolia, it is not surprising that most of the personnel would be Han Chinese. I do not mean to give the impression that the majority of grassland scientists working throughout Inner Mongolia are Han. The provincial capital of Hohhot boasts two of the top national universities—the University of Forestry and the University of Agriculture and Animal Husbandry—that produce substantial numbers of Mongol students who specialize in grassland studies and become administrative cadres throughout the region (Uradyn Bulag, personal communication).

7. When social research is conducted by foreigners in China, the host institution is usually affiliated either with the Chinese Academy of Social Sciences or a local university. In rural settings, Chinese research assistants are often assigned to escort and intervene, should the scholar stray from an approved agenda stipulating potential destinations and interview questions. Local officials from the Public Security Bureau are also prone to run interference in the case of "sensitive" research. Since my host institution was affiliated with the Chinese Academy of Sciences, they had no clear operating procedures to monitor the work of social scientists. This permitted me an unusual degree of autonomy, especially in the field setting, where I was free to work under my own discretion, often for weeks at a time without any of the scientific staff being present.

8. Typical soil and dune fixation experiments involve planting a variety of drought resistant tree, shrub, and grass species on different types of enclosed soils and dunes to record the biological and geophysical results over time (growth of biomass, root length, composition of vegetation, changes in soil pH, soil carbon levels, distance of dune advance, etc.). Experiments in meadow salinization involve variations in the application of gypsum, in the selection and methods of planting salt-resistant grasses, and in the diversion of flood water. The experimental plots are intended to demonstrate to the community both the technical procedures and the economic rationale for replacing traditional and extensive management strategies with a more labor- and capital-intensive husbandry.

9. Throughout the book, I use pinyin Chinese (instead of Mongolian) for local terminology. This is consistent with the fact that I conducted research in China using the Chinese language (within a bilingual community), and that my analysis interprets land use changes in the context of the post-Mao reform era.

10. The *kang* is a raised platform utilized as a bed that is internally heated through vents that connect to the cooking hearth.

11. The most basic concerns we shared were to help control extensive soil erosion in the region, to help alleviate widespread poverty, and to help facilitate communication between officials, scientists, and local residents. The process of working toward these common goals made us allies more often than not. For example, we worked together to overcome many logistical and institutional barriers to make my visit possible in the first place. Beyond that, severe weather and the general impoverishment of the region frequently imposed emergency conditions upon us that further strengthened our mutual camaraderie. We traveled together (perilously) through floods, snowstorms, dust storms, and mudslides; we shared personal resources like food, gasoline, batteries, medicines, toilet paper, and even sleeping quarters; and we entertained each other through sports, stories, songs, and the consumption of alcohol.

12. My health and life were threatened on many occasions by circumstances as diverse as knife-fighting, alcohol poisoning, dog attack, horse-riding accidents, third-degree burns, and exposure.

## Chapter 2: Land Degradation and the Chinese Discourse

1. Not only does soil fertility and stability vary from one location to the next, but plant species are also quite diverse. Longworth and Williamson (1993: 81) indicate that over 900 different plant species grow on the rangeland of Inner Mongolia, but only 210 species are grazed by livestock.

2. There are several historical complexities that such broad generalizations do not take into account. One is the fact that the territorial boundaries of IMAR have not been static through time. Another is the fact that since 1979, many residents who formerly registered as Han began to change their ethnic classification to Mongol in order to take advantage of new privileges for national minorities, especially exemptions from the one-child policy.

3. Many readers may be aware that Kevin Stuart has written a critical evaluation of Lattimore's travelogues in his book *Mongols in Western/American Consciousness* (1997). I consider it worthwhile to point out that Stuart's strident commentary is concerned with Lattimore's professional ethics and personality traits, and not with his regional knowledge or reliability on technical matters. Stuart himself emphasizes that "many Mongols did respond positively to Lattimore because of his knowledge of the Mongol language, history, and culture" (137). Where I quote Lattimore, I consider him a credible and widely accessible source of historical information for the region under study.

4. Mao eventually called off the campaign, but the bitter memories and ecological consequences linger. Despite the ultimate policy reversal, the political fervor for grain production always looms as a potential threat in the minds of grassland officials and residents.

5. *Pohuai* generally means "to destroy." In this case, the Xinhua news agency rendered the translation as "sabotage."

6. For example, a senior Chinese scientist and vice minister of Science and Technology publicly complained that "the irrational and plundering exploitation of natural resources is very serious in China" (Deng Nan, quoted in *China Daily* 1992). In 1990, a group of deputies to The Seventh National People's Congress drafted a letter to the Standing Committee and its chairman that proposed the formation of a new environmental committee to protect national resources. The letter charged that "improper use, waste, and destruction of resources can be found everywhere," but explicitly related the problems of soil erosion, desertification, and degeneration of grassland to patterns of mismanagement by local residents (Xinhua 1990).

7. Also consider the many assertions by Chinese journalists of the Maoist era who were assigned to report the tremendous transformations of nature that collectivization had made possible. One early article proclaimed: "The battle of the desert had decisively turned. The old pattern of men retreating before it was a thing of the past; whereas once the sands had crept up at the rate of three to eight metres annually, today men marched at twenty kilometres a year into its very heart and subjected it to their will" (Hsiao 1960: 25).

8. The journalists make the point that a hundred years earlier, the pastures of Nasihan attracted two Mongol nobles who brought their herds and their serf labor to graze the area. But they were unsuccessful in their attempts to draw water, so they vacated the "wasteland." In 1959, however, the collective dynamited underground rock ("at the very same spot where the two lords had failed") to create a reliable well that permitted prosperous settlement. "Four years later, land which had been nothing but shifting sand dunes had changed to luxuriant grazing grounds covered with thick grass and low bushes." The authors further claim that reserves of winter fodder were so abundant that "bad weather no longer constitutes a threat to the herds," and conclude with a resounding affirmation that, thanks to communal living, "prosperity had come to the grasslands."

9. The Sanbei Shelterbelt Region encompasses 551 counties in north, northwest, and northeast China. The reforestation project is expected to take more than seventy years. In the end, the afforested area is to cover about thirty-five million hectares.

10. No doubt, after decades of neglect, the problem of desert expansion has gained more prominent attention, at least rhetorically, since the beginning of the reform era. Officials and scholars routinely declare it to be a national "top priority" (Xu Youfang 1993; Wang, Wang, and Zhang 1993: 10).

11. The China State Council has set up a National Coordination Panel for Desertification Control to assume responsibility for a unified plan of attack. The panel consists of officials selected from all the following agencies: National Afforestation Committee, Ministry of Forestry, Ministry of Water Conservancy, Ministry of Agriculture, Chinese Academy of Sciences, State Planning Commission, Ministry of Finance, Ministry of Energy, Ministry of Railways, State Science and Technology Commission, National Environment Protection Agency, State Administration of Land Management, State Administration of Taxation, People's

Bank of China, Office for Promoting Economic Development in Poverty Areas, and the Office of National Agricultural Integrated Development (China State Council 1994: 182; Wang, Wang, and Zhang 1993: 10). The reform government's emphasis on programmatic response has manifested itself in other new directions as well. For example, since 1991, China has convened five biannual national conferences for the control of desertification.

## Chapter 3: The Ambiguities of Land Degradation

1. *Tuihua* is the opposite of *jinhua*, a term connoting evolutionary advancement in both a biological and social sense. Within Chinese Communist ideology, *tuihua* conveys a decidedly negative moral resonance. In theory, the historical march toward communism should be characterized by continuous *jinhua*, so the presence of *tuihua* indicates an interruption that runs contrary to the flow of history. (Thanks to Stevan Harrell for sharing this insight.)

2. Given that some land system transformations are ephemeral and all are reversible over a sufficiently long time frame, the attempt to distinguish between "temporary" and "permanent" changes in the land can be an arbitrary exercise. The determination is emphatically dependent upon shifting subjective priorities. Comparable difficulties prevent the possibility of drawing clear distinctions between "natural" forces and "human" interference, or between "significant" and "tolerable" transformations in the land. Other studies have highlighted similar definitional problems with the term "productivity." For example, different assessments of land quality may result depending upon how and when biological productivity is measured—whether in terms of caloric content or protein, and whether during critical moments of the season or over the course of an annual cycle. Some scholars have demonstrated that even rangeland transition toward healthy but unpalatable plant communities may qualify as degradation according to narrow interpretations of productivity (see Conant 1982; Breman et al. 1979).

3. According to a report of the China National Committee of the International Decade for Natural Disaster Reduction (IDNDR 1993: 9–14), central and local governments are allocating money for long-term disaster prevention measures through the 1990s in the following order of priority (and estimated cost): comprehensive water conservancy (102.4 billion yuan); shelter-forest engineering and fire prevention (30 billion yuan); seismic monitoring, resistance, and prevention (10.1 billion yuan); grassland management, including control of rodents, insects, and fires (6.1 billion yuan); national desertification control and prevention (4.5 billion yuan); marine disaster forecasting and warning (200 million yuan); monitoring for pestilence and crop disease (180 million yuan); and landslide control (57 million yuan).

4. For a review of these various changes, see Longworth and Williamson 1993: 42–47.

5. In Mandarin, the phonemes for "east" (*dong*) and "west" (*xi*) are combined to form the word meaning "thing" (*dongxi*). The negative term *mei* is then added to form the common expression "nothing" (*mei dongxi*). Those people who are neither east nor west often joke that their relationship to the government is quite literally "nothing."

### Chapter 4: The Land in Cultural Context

1. Such broad analytical categories necessarily involve some generalizations that conceal underlying variation and nonconformity. For example, while there may not be absolute consensus among all urban Han Chinese and nationalist intellectuals regarding the grassland environment, there is nevertheless a conventional view that can be clearly discerned in higher-profile public writings and spoken comments. I believe the same argument holds for the other idealized types presented here. Note especially that this study is concerned with pastoral Mongols living in the countryside; I make no representations about the views of urban Mongols living in China.

2. The popular legend of Meng-Jiang Nü testifies to the counter-motif of tyranny associated with the Great Wall. She was a courageous "widow of the Wall," and her story, as celebrated in opera and fable, includes the following highlights: Her husband was an honest scholar before soldiers conscripted him on his wedding day to build a section of the Great Wall. Like hundreds of thousands of workers before him, he died in the labor, and his body was incorporated into the foundation of the Wall. Informed by a dream, Meng-Jiang Nü then journeyed to the site and, with her head, smashed a section of wall and entered it to find an endless pile of bones. She bit her finger to drip blood on the skeletons. The blood rolled off all the bones except one pile, which soaked up her blood like a sponge. Sure that she had found the remains of her husband, she collected and arranged them for a proper burial. She was arrested by the emperor, but denounced him as a foolish and self-indulgent ruler. The emperor was impressed by her beauty and fidelity, and he asked her to join his harem. She agreed on three conditions: a gold coffin had to be provided for the bones of her husband, all court officials had to go into mourning, and the emperor himself had to march in the funeral procession. Once these conditions were met, she then committed suicide (Cheng 1984: 44–50).

3. This cultural perception is not necessarily accurate. Mongol herders have historically manipulated or managed the appearance and structure of "natural" rangeland through a variety of techniques such as selective grazing, mowing, fallow land succession, and anthropogenic fire.

4. I entertained elaborate discussions over the year about concerns such as meal preparations or housing and travel arrangements, but my hosts asked very little about the nature or the progress of my work.

5. I can only relate the circumstances of my own research experience in China. Participants of other collaborative efforts can speak for themselves. I intend no specific evaluation of their own successes or failures.

### Chapter 5: The Community

1. This boundary has been fluid through history, and the tidy imagery of a sharp and lasting polarity between nomads and farmers is increasingly under attack (see Di Cosmo 1994).

2. According to Lattimore, this was done to prevent Mongols from settling on the land. To find legal precedent for this outright theft of territory from Mongol farmers, authorities relied upon Qing Dynasty edicts of 1748, which actually intended to check agricultural expansion in the grasslands. The edicts forbade Mongol tribal princes from allotting land for cultivation to either Han or Mon-

gols without direct confirmation from local authorities who supervised Mongol affairs on behalf of the imperial Qing government. Lattimore (1934: 105) was indignant at the transparently unjust interpretation of the outdated edicts: "To construe them afresh as establishing the legal principle that Mongol land does not belong to the Mongols but to the Chinese state, and to apply this principle for the purpose of expropriating land from the Mongols, whether or not already farmed, and preventing the Mongols from saving themselves by taking up agriculture, appears to me to be a masterly example of legal chicane and historical cynicism."

3. The supply of electricity remains highly dependent upon fair weather. Strong winds, low temperatures, or even rain can interrupt the power supply for days at a time. My journal indicates that electricity failed for more than forty days while I was physically present in Wulanaodu.

4. The national survey sampled Xianghuang banner in Xilingele, Inner Mongolia, where ethnic Mongols comprise more than 50 percent of the population.

5. One of my closest informants communicated this to me bluntly with a clever pun: "*meiyou yan, meiyou jiu, bu neng yanjiu*" (Without cigarettes or liquor, you cannot do research). In Mandarin, the phonemes meaning "research" can be broken down into separate phonemes that signify both "cigarette" and "liquor."

6. The fact that women do sometimes indulge in alcohol consumption does not invalidate the "hypermasculinity" that drinking confers upon males. On the contrary, female drinking (like smoking) seems to fortify both local and national perceptions that Mongol women are also more robust and "virile" than Han women.

7. It is certainly possible that the female population has its own (gender-specific) destructive dynamic with regard to alcohol consumption, but I found little evidence to support that view. That said, a foreign male visitor to the community has rather minimal opportunity to interact with or inquire about the private lives of women.

## Chapter 6: Enclosure and Changes in Physical Landscape

1. The reserved field, which has since expanded greatly, appears red in sharp relief at the center of Figure 5.1.

2. China is and remains a socialist state, so that land technically belongs to the people's government. Decollectivization initiated the redistribution to private households of land use rights (usufruct) only.

3. A notable exception is found in a study by Liu (1990: 97–100), who explicitly recognized the relationship between expanding enclosures and grazing pressure intensification in surrounding grasslands.

4. Technically speaking, the 1984 distribution of rangeland resources allocated user rights only to *lianhu*, but in reality, many of these household cluster groups functioned independently from the start. By 1988, discord was so prevalent that local authorities conceded to a second distribution that formally divided the range among independent households.

5. I collected enclosure data and detailed herding strategies in Wulanaodu village through 154 lengthy household interviews, while livestock data came from annual village registries. Combining these data sets, I then prepared a database

for the entire village and computed how many total cows, sheep, goats, horses, donkeys, and camels in the village grazed outside all private enclosures during each season of each year.

Figures for Tables 6.1 and 6.2 are based on a total land area defined as 100,000 mu, or 6,670 hectares (1 mu = 0.067 hectares). The most recent land survey estimates the total land area of Wulanaodu at 132,914 mu, but roughly 33,000 mu must be subtracted to account for unusable moving dunes (22,000 mu), collectively enclosed forested area that is never grazed (10,200 mu), land reserved for cultivation of fodder crops (600 mu), and unenclosed pond area (200 mu). Following the Chinese convention as outlined in the Chifeng Grassland Management Rules (Chifengshi caoyuan jianlisuo 1990: 217), one SEU is defined here as any combination of animals with a total forage demand roughly equal to that of one adult ewe, or 2.4 kilograms of dry matter per day. Each cow counts as 5 SEUs based on an assumed forage demand substitution ratio of 5, or 12.0 kg. of dry matter per day. Each goat counts as 0.9, each horse as 6, each donkey as 3, and each camel as 7 SEUs. Finally, I should make explicit that I am juxtaposing newly structured patterns of grazing against the pre-reform grazing pattern in which collective herds roamed across collective rangeland, so that grazing pressure was rather evenly distributed.

Regarding the use of annual village registries, my analysis is uneasily dependent upon the reliability of government records. I do not hold a naive assumption that all reported numbers are absolutely accurate, but I do accept the estimate by numerous official and civilian sources that annual household livestock figures are at least 90 percent accurate. This estimate conforms with my own experiences in checking reported figures against personal household records whenever they were available. Informants tell me that surprise livestock audits do occur often enough (especially since 1987) to keep the reported figures fairly close to reality. In any case, reporting errors would most likely produce an underestimation of livestock, since residents would attempt to reduce rather than increase their tax liability. If anything, the numbers err on the side of conservation, which means that the troubling SEU/hectare ratios for the summer-fall growing season would only be higher than reported in Table 6.2.

## Chapter 7: Enclosure and Changes in Social Landscape

1. This measure takes into account not only the relative size of enclosed land, but also the benefits derived from enclosing earlier than others. It is derived by multiplying the size of each separate enclosure by the number of years it has been enclosed, summing the total for each household, then dividing by fifteen (the number of total years possible).

2. According to official estimates, each sheep or goat requires 7 mu of rangeland, and each cow requires 35 mu. Therefore, 700 mu of rangeland should provide for the nutritional requirements of 100 sheep or goats, or 20 cows, or any equivalent mixture thereof. Households exceeding that ratio quota are supposed to be fined, although no fines have ever been levied in Nasihan since decollectivization.

3. The ancient art of geomancy essentially holds that no two positions on the

earth have equal value, although the relative rank of each may well be a matter of subjective opinion (Bruun 1995: 183).

4. Actually, hypothermia is a widely underestimated and underreported health problem, whether in China or the United States (Lloyd 1986: 3). Government statistics would therefore not be likely to provide a credible measure for the phenomenon under study. In any case, the relevant social and physical realities of Nasihan may well be so geographically limited that even if all the cases of cold-related injury and death were accurately recorded, the meaningful numbers would be drowned in a sea of rural population statistics.

5. The traditional alcoholic beverage was kumiss (known as *airag* in Mongolian and *suan ma nai* in Chinese). It was a mild home product derived from the fermented milk of a female horse that generally contained no more than 2 percent alcohol (Montell 1937: 322). Kumiss can be converted into more potent alcoholic beverages depending upon the number of distillations. In Chinese, these beverages are indiscriminately called *naijiu* (milk liquor), whereas Mongolian employs distinct terms for different potencies (such as *arhi* and *araza*).

6. Smoking, which is nearly universal in Nasihan among both males and females, has a similar effect upon appetite.

7. The power of this experiential truth is artistically evoked in the ethnographic film *Taiga* (Ottinger 1992), which temporarily reproduces the expansive scale and plodding pace of the traditional Mongolian lifestyle for urban audiences by treating them to 500 minutes of visual documentation.

## Chapter 8: Landscape and Identity

1. Vandalism can render an enclosure dysfunctional until someone repairs the wire, usually within a few days' time. While local fences can be rather permeable, they have sufficient integrity to structure the grazing patterns described earlier. I do not mean to imply that they are useless.

2. To provide an initial frame of reference, I should briefly describe the different landscapes from my own subjective perspective. Figure 8.1 (the "best") reveals a dense and unbroken field of tall grasses and shrubs that are edible to livestock. The abundant forage suggests the land is utilized as a reserve meadow for collective hay production. Figure 8.2 (the "worst") reveals a shallow pond that has formed on terrain dominated by mobile sand dunes. Withered shrub roots gnawed by passing livestock provide the only sign of vegetation.

3. It is relevant to note that some high-profile research conducted in China on the fringes of the Taklimakan Desert, along the Hexi Corridor, and in Turpan have reported significant gains in vegetative cover (from 60 to 85 percent) on enclosed sandy lands within three years' time. Those experiments, however, involved intensive labor and capital outputs, as well as winter irrigation (see DDR 1982: 16). The land was not simply enclosed and fallowed, as in the hypothetical scenario that I used to query my respondents. Humphrey and Sneath (1999: 106) have reported that grassland specialists in Xinjiang estimate that it takes fifteen to twenty years for ploughed land to return to its previous productivity as pasture.

4. Again, I should divulge my own subjective interpretations of these photos.

Figure 8.3 reveals a diverse steppe terrain that includes *Caragana* shrub, elm trees with exposed roots, diffuse and light grass cover, and the pervasive presence of wind-blown sand. The composition suggests the area is utilized as public range. Figure 8.4 reveals a dense and homogenous field of tall grass, undoubtedly utilized as a reserve hay field. Figure 8.5 reveals a sand-covered terrain dotted with *Caragana* shrubs and elm trees, whose foliage has been heavily grazed by passing livestock.

5. This figure derives from the following calculations: In 1993, Wulanaodu household livestock registries indicated a total of 1,270 sheep and 4,012 goats in the village. I estimated that each goat would produce an average of 0.5 kg of cashmere (some more, some less, depending upon the breed). Thus, about 2,000 kg of cashmere were produced, which was used to hide between 400 and 500 kg of sand (20–25 percent). Improved breeds of sheep are expected to produce 15–20 kg of wool each, but unimproved breeds yield only half that much. The high and low sheep yield for the entire village would therefore range between 19,050 and 9,525 kg of wool, which would hide between 952 and 476 kg of sand (5 percent). I added the low estimates for both cashmere and wool (400 + 476) to determine that at least 876 kg of sand were marketed. I added the high estimates (500 + 952) to determine that no more than 1,450 kg of sand were marketed. I took the average of these two figures to conclude that roughly 1,160 kg of sand were sold in 1993.

6. This range was derived using the figures provided by Longworth and Williamson (1993: 31), who reported a total of 2,076 tons of cashmere, 2,292 tons of goat wool, and 59,203 tons of sheep wool produced and sold in IMAR for the year 1990. At 5 percent adulteration in wool, that amount would yield about 3,075 tons of sand. At 20–25 percent adulteration in cashmere, that amount would yield between 415 and 519 tons of sand. The estimate from wool was then added to the high and low estimates from cashmere to produce the final range reported.

7. The processing plants in Chifeng complain every year about how difficult it is to scour and clean the hair and wool. Apparently the practices have become even more unscrupulous in some areas of the country since decollectivization. Pieces of iron and glass are now falling from the produce on some factory floors, and adulteration at that level cannot be amended.

8. This suggests that at least some members of the scientific community were engaged in explicitly illegal and immoral practices. They were essentially gambling with government finances for secretive private profits, while diverting their resources and their labor from the needs of the community in order to compete with local herders on the market.

9. It is reasonable to conclude that many local residents currently engage in a variety of activities that appear destructive to the grassland ecosystem. These activities, however, are forms of cultural and political resistance to an environmental regime imposed upon the community from outside. The destruction is driven not by deficiencies of tradition (as the national discourse would have it), but by poorly managed transition in the wake of rangeland privatization.

# Bibliography

Abbey, Edward
   1984   Desert Images. In *Beyond the Wall: Essays from the Outside*, pp. 77–94. New York: Holt, Rinehart and Winston.

Aird, John
   1990   *Slaughter of the Innocents: Coercive Birth Control in China.* Washington, D.C.: American Enterprise Institute Press.

Anderson, Benedict
   1983   *Imagined Communities: Reflections on the Origin and Spread of Nationalism.* London: Verso.

Asia Watch
   1992   *Continuing Crackdown in Inner Mongolia: An Asia Watch Report.* New York: Author.

Associated Press (AP)
   1994   Nature Finds an Independent Ally. *South China Morning Post* (Hong Kong), April 2.

Ba Gen and the Chifeng Grassland Work Station
   1993   Jiating xuqun xiaocao kulun jianshe qianyi [Overview of Householder Herd and Grassland Construction Policy]. *Nei Menggu Cao Yie* 2: 19–22.

Barfield, Thomas J.
   1993   *The Nomadic Alternative.* Englewood Cliffs, N.J.: Prentice Hall.
   1989   *The Perilous Frontier: Nomadic Empires and China.* Oxford: Basil Blackwell.

Barth, Frederic
   1966   Introduction to *Models of Social Organization*, pp. 9–38. Occasional Papers of the Royal Anthropological Institute, no. 23. London: Royal Anthropological Institute.

Bawden, Charles
   1968   *The Modern History of Mongolia.* New York: Frederick A. Praeger.

Becker, Jasper
   1992   *The Lost Country: Mongolia Revealed.* London: Hodder and Stoughton.

*Beijing Review*
   1988    Prices Raised for Cigarettes, Liquor. August 1–7, p. 7.
Bessac, Frank
   1965    Revolution and Government in Inner Mongolia: 1945–50. *Papers of the Michigan Academy of Science, Arts, and Letters* 50: 415–429.
Blaikie, Piers
   1989    Explanation and Policy in Land Degradation and Rehabilitation for Developing Countries. *Land Degradation and Rehabilitation* 1(1): 23–37.
   1985    *The Political Economy of Soil Erosion in Developing Countries*. New York: Longman.
Blaikie, Piers, and Harold Brookfield
   1987    *Land Degradation and Society*. New York: Methuen.
BMOF, *see* Bureau of the Ministry of Forestry
Bocking, Stephen
   1994    Visions of Nature and Society: A History of the Ecosystem Concept. *Alternatives* 20(3): 12–18.
Bohm, David, and F. David Peat
   1987    *Science, Order, and Creativity*. New York: Bantam Books.
Boserup, Ester
   1965    *The Conditions of Agricultural Growth*. New York: Aldine.
Boxer, Baruch
   1991    China's Environment: Issues and Economic Implications. In *China's Economic Dilemmas in the 1990's: The Problems of Reforms, Modernization, and Interdependence*, submitted to the Joint Economic Committee Congress of the United States, vol. 1 (April): 290–307.
Breman, H., et al.
   1979    Pasture Dynamics and Forage Availability in the Sahel. *Israel Journal of Botany* 28: 227–251.
Bromley, Daniel, and Michael Cernea
   1989    *The Management of Common Property Natural Resources: Some Conceptual and Operational Fallacies*. Washington, D.C.: World Bank.
Broughton, Douglas
   1947    *Mongolian Plains and Japanese Prisons*. London: Pickering and Inglis.
Brown, Colin, and John Longworth
   1992    Multilateral Assistance and Sustainable Development: The Case of an IFAD Project in the Pastoral Region of China. *World Development* 20(11): 1663–1674.
Bruun, Ole
   1995    Fengshui and the Chinese Perception of Nature. In *Asian Perceptions of Nature*, ed. Ole Bruun and Arne Kalland, pp. 173–188. Nordic Institute of Asian Studies, no. 18. Great Britain: Curzon Press.
Bruun, Ole, and Ole Odgaard, eds.
   1996    *Mongolia in Transition*. Surrey: Curzon Press.

Bureau of the Ministry of Forestry, China (BMOF)
1990    Spring Has Come to the Desert. In *A Rising Green Great Wall: Construction of the Sanbei Shelter-Forest System*, ed. Bureau of the Ministry of Forestry in charge of construction of shelter forest in North, Northeast, and Northwest China. Beijing: Da Di Publishing House.

Buzdar, Nek
1992    The Role of Institutions in the Management of Commonly-Owned Rangelands in Baluchistan. In *Sociology of Natural Resources*, ed. Michael Dove and Carol Carpenter, pp. 218–238. Karachi, Pakistan: Vanguard Press.

Cao Xinsun et al.
1984    Wulanaodu diqu shengtai xitong de jiegou gongneng yu gaishan tujing [The Structure, Function, and Restorative Methods for the Wulanaodu Ecosystem]. In *Fengsha ganhan zonghe zhili yanjiu: Neimenggu dongbu dichu* [Studies on the Integrated Control of Wind, Sand Drifting, and Drought in Eastern Inner Mongolia], ed. Cao Xinsun, vol. 1, pp. 1–7. Hohhot: Inner Mongolia People's Publishing House.

Carpenter, Robert A.
1989    Giant Panda Controversy. *Buzzworm* 1(2): 16–29.

Castells, Manuel, and Jeffrey Henderson
1987    Introduction to *Global Restructuring and Territorial Development*, ed. Jeffrey Henderson and Manuel Castells, pp. 1–17. London: Sage.

CCP, *see* Chinese Communist Party

Chang Sen-dou
1977    The Morphology of Walled Capitals. In *The City in Late Imperial China*, ed. G. William Skinner, pp. 75–100. Stanford, Calif.: Stanford University Press.

Chang Xiaochuan, Cai Weiqi, and Xu Qi
1990    Effects of Fencing on Soil-Vegetation Systems and the Use of Grasslands by Herbivorous Animals. In *International Symposium on Grassland Vegetation*, pp. 461–468. Beijing: Science Press.

Chen Junshi et al.
1990    *Diet, Life-style, and Mortality in China: A Study of the Characteristics of 65 Chinese Counties*. Ithaca, N.Y.: Cornell University Press.

Cheng Dalin
1984    *The Great Wall of China*. Hong Kong: South China Morning Post Ltd.

Chevrier, Yves
1988    NEP and Beyond: The Transition to "Modernization" in China. In *Transforming China's Economy in the Eighties*, ed. S. Feuchtwang, A. Hussain, and T. Pairault, pp. 7–35. Boulder, Colo.: Westview Press.

Chifengshi caoyuan jianlisuo [Chifeng City Grassland Regulatory Office]
1990    *Caodi guanli fagui yu guicheng ziliao huibian* [Collection of Grassland Management Rules and Regulations]. Chifeng, Inner Mongolia Autonomous Region.

*China Daily*
1994   Income Gap Stimulates Growth. July 1.
1992   Poor Environment a Threat. May 8.
1991   Stop the Deserts. July 31 (excerpted from *Renmin ribao*, July 29).
1988   China Seeds Desert by Plane. August 10.

China State Council
1994   *China's Agenda 21: White Paper on China's Population, Environment, and Development in the 21st Century.* (Adopted at the 16th Executive Meeting of the State Council, March 25, 1994.) Beijing: China Environmental Science Press.

Chinese Communist Party (CCP)
1981   *The Sixth Five-Year Plan of the People's Republic of China for Economic and Social Development, 1981–85.* Beijing: Foreign Language Press.
1978   Communique of the Third Plenary Session of the 11th Central Committee. *Beijing Review* 21(52): 6–16.

Chonghalakoushu and Jisizhengli
1986   Songshu shan [Pine Mountain]. In *Wengniuteqi wenshi ziliao* [Historical Materials on Wengniute Banner], pp. 103–105. Wudan: Wengniute weisheng xitong yinshaochang.

Chuluun Togtohyn, Dennis Ojima, Jargalsaihan Luvsandorjiin, Jerrold Dodd, and Stephen Williams
1993   Simulation Studies of Grazing in Mongolian Grassland Ecosystems. Paper prepared for the research conference Grassland Ecosystem of the Mongolian Steppe, Racine, Wisconsin, November 4–7.

Cohen, Anthony
1985   *The Symbolic Construction of Community.* New York: Tavistock Publications.

Conant, Francis P.
1982   Thorns Paired, Sharply Recurved: Cultural Controls and Rangeland Quality in East Africa. In *Desertification and Development: Dryland Ecology in Social Perspective*, ed. Brian Spooner and H. S. Mann, pp. 111–122. New York: Academic Press.

Connor, Walker
1984   *The National Question in Marxist-Leninist Theory and Strategy.* Princeton, N.J.: Princeton University Press.

Corbett, Jim
1991   *Goatwalking.* New York: Viking Press.

Cousins, B., D. Weiner, and N. Amin
1992   Social Differentiation in the Communal Lands of Zimbabwe. *Review of African Political Economy* 53: 5–24.

Croll, Elisabeth, Delia Davin, and Penny Kan, eds.
1985   *China's One-Child Policy.* New York: St. Martin's Press.

Cronon, William
  1983    *Changes in the Land: Indians, Colonists, and the Ecology of New England.* New York: Hill and Wang.
Crook, Frederick
  1990    Allocation of Crop Sown Area: Analysis of Trends and Outlook for the Future. *China Agriculture and Trade Report,* July, pp. 37ff. U.S. Department of Agriculture.
Crumley, Carole
  1994    Historical Ecology: A Multidimensional Ecological Orientation. In *Historical Ecology: Cultural Knowledge and Changing Landscapes,* ed. Carole Crumley, pp. 1–16. Santa Fe, N.M.: School of American Research Press.
Dai Qing
  1994    *Yangtze! Yangtze! Debate over the Three Gorges Project.* Trans. Nancy Liu et al. London: Earthscan.
DDR, *see* Department of Desert Research
Deal, David
  1984    The Question of Nationalities in Twentieth-Century China. *Journal of Ethnic Studies* 12(3): 23–53.
Department of Desert Research (DDR; also Lanzhou Institute of Desert Research)
  1982    The Transformation of Deserts in China: A Summary View of the People's Experiences in Controlling Sand. In *Combating Desertification in China,* ed. James Walls. Nairobi: United Nations Environment Programme.
De Riencourt, Amaury
  1958    *The Soul of China.* London: Honeyglen.
Di Cosmo, Nicola
  1994    Ancient Inner Asian Nomads: Their Economic Basis and Its Significance in Chinese History. *Journal of Asian Studies* 53(4): 1092–1126.
Dove, Michael R.
  1998    Privileged Ecotypes in Southeast Asia: Ecological Models, Authority, and Bias in Environmental Representation. In *Representing Natural Resource Development in Asia: "Modern" Versus "Postmodern" Scholarly Authority,* ed. Michael Fischer, pp. 1–22. Centre for Social Anthropology and Computing Monographs, CSAC Human Ecology Series 1, University of Kent at Canterbury. Also available on the Internet at http://lucy.ukc.ac.uk/Postmodern/Michael_Dove_TOC.html.
  1986    Peasant Versus Government Perception and Use of the Environment: A Case Study of Banjarese Ecology and River Basin Development in South Kalimantan. *Journal of Southeast Asian Studies* 17(1): 113–136.
Dregne, H. E.
  1983    *Desertification of Arid Lands.* New York: Harwood Academic Publishers.

Engel, J. Ronald
  1983   *Sacred Sands: The Struggle for Community in the Indiana Dunes.*
          Middletown, Conn.: Wesleyan University Press.

Feeny, David, Fikret Berkes, Bonnie McCay, and James Acheson
  1990   The Tragedy of the Commons: Twenty-Two Years Later. *Human Ecol-*
          *ogy* 18(1): 1–19.

Fei Xiaotong
  1984   Bianqu kaifa Chifeng pian [Essay on border development in Chifeng].
          *Neimeng ribao*, March 11.

Fernandez-Gimenez, Maria
  1995   Mobility, Reciprocity and Local Ecological Knowledge: The Basis for
          Sustainability on the Mongolian Steppe? Paper presented at the annual
          meeting of the Association for Asian Studies, Washington, D.C., April
          6.

Foucault, Michel
  1980   *Power/Knowledge: Selected Interviews and Other Writings,*
          *1972–1977*, ed. and trans. Colin Gordon. New York: Pantheon.
  1979   *Discipline and Punish: The Birth of the Prison*, trans. Alan Sheridan.
          New York: Vintage.

Frank, David John
  1997   Science, Nature, and the Globalization of the Environment,
          1870–1990. *Social Forces* 76(2): 409–437.

Fratkin, Elliot
  1997   Pastoralism: Governance and Development Issues. *Annual Review of*
          *Anthropology* 26: 235–261.

Fullen, Michael, and David Mitchell
  1991   Taming the Shamo Dragon. *Geographical Magazine*, November 26.

Gilley, Bruce
  2000   Saving the West. *Far Eastern Economic Review*, May 4: 22–23.

Gilmour, Rev. James
  1883[?] *Among the Mongols.* London: The Religious Tract Society.

Gladney, Dru
  1991   *Muslim Chinese.* Cambridge, Mass.: Harvard University Press.

Goldstein, Melvyn, Cynthia Beall, and R. P. Cincotta
  1990   Traditional Conservation on Tibet's Northern Plateau. *National Geo-*
          *graphic Research* 6(2): 139–156.

Grapard, Allan
  1994   Geosophia, Geognosis, and Geopiety. In *NowHere: Space, Time, and*
          *Modernity*, ed. Roger Friedland and Deirdre Boden, pp. 372–401.
          Berkeley and Los Angeles: University of California Press.

Grousset, Rene
  1967   *Conqueror of the World.* Trans. Denis Sinor and Marian Mackellar.
          London: Oliver and Boyd.

Guo Yong (vice governor [*fu qi zhang*] of Wengniute banner)
  1993   Personal interview with author, Wulanaodu, September 3.

Haas, Peter M.
  1992   Introduction: Epistemic Communities and International Policy Coordination. *International Organization* 46(1): 1–35.
  1989   Do Regimes Matter? Epistemic Communities and Mediterranean Pollution Control. *International Organization* 43(3): 377–403.
Hardin, Garrett
  1968   The Tragedy of the Commons. *Science* 162: 1243–1248.
Harrell, Stevan
  1995   Introduction: Civilizing Projects and the Reaction to Them. In *Cultural Encounters on China's Ethnic Frontiers*, ed. Stevan Harrell, pp. 3–36. Seattle: University of Washington Press.
Harvey, David
  1990   Between Space and Time: Reflections on the Geographical Imagination. *Annals of the Association of American Geographers* 80(3): 418–434.
He Bochuan
  1991   *China on the Edge: The Crisis of Ecology and Development.* San Francisco: China Books and Periodicals.
Heathcote, R. L.
  1980   Summary and Conclusions: The Role of Perception in the Desertification Process. In *Perception of Desertification*, ed. R. L. Heathcote, pp. 120–134. Tokyo: United Nations University.
  1983   *The Arid Lands: Their Use and Abuse.* New York: Longman.
Hedley, Rev. John
  1910   *Tramps in Dark Mongolia.* New York: Charles Scribner's Sons.
  1906   *On Tramp Among the Mongols.* Rpt. from the *North China Daily News.* Taipei, Taiwan: Cheng Wen Publishing Company.
Hinton, William
  1990   *The Great Reversal: The Privatization of China, 1978–1989.* New York: Monthly Review Press.
Hjort, Anders
  1982   A Critique of Ecological Models of Pastoral Land Use. *Nomadic Peoples* 10: 11–27.
Ho, Peter
  1998   Ownership and Control in Chinese Rangeland Management Since Mao: A Case Study of the Free-Rider Problem in Pastoral Areas in Ningxia. In *Cooperative and Collective in China's Rural Development*, ed. Eduard B. Vermeer, Frank N. Pieke, and Woei Lien Chong, pp. 196–235. Armonk, N.Y.: M. E. Sharpe.
Ho, Samuel, and Ralph Huenemann
  1984   *China's Open Door Policy: The Quest for Foreign Technology and Capital.* Vancouver: University of British Columbia Press.
Honey, David
  1992   *Stripping Off Felt and Fur: An Essay on Nomadic Sinification.* Papers

on Inner Asia, no. 21. Bloomington, Ind.: Research Institute for Inner Asian Studies.

Howard, Pat
  1988   *Breaking the Iron Rice Bowl: Prospects for Socialism in China's Countryside*. Armonk, N.Y.: M. E. Sharpe.

Hsiao Ming
  1960   Pushing Back the Deserts of Gansu. *Beijing Review* 29 (July 19): 24–25.

Hu Mingge (deputy director of the Animal Husbandry Modernization Office, Chifeng City)
  1994   Personal interview with author, Chifeng City, IMAR, June.

Hu Mingge, ed.
  1990   *Chifeng caodi* [Chifeng Grasslands]. Beijing: Agricultural Press.

Hua Guofeng
  1978   Unite and Strive to Build a Modern, Powerful Socialist Country. *Beijing Review* 21(10): 7–40.

Huang Wenxiu
  1989   Woguo caodi xumuye de fazhan qianjing yu tujing [The Prospect and Pathway for Pastoral Animal Husbandry Development in China]. In *Zhongguo caodi kexue yu caoye fazhan* [Development of Grassland Science and Prataculture in China], pp. 42–46. Beijing: Science Press.

Huffman, James
  1986   China. In *Government Liability and Disaster Mitigation*, pp. 23–80. Boston: University Press of America.

Humphrey, Caroline, and David Sneath, eds.
  1999   *End of Nomadism? Society, State, and the Environment in Inner Asia.* Durham, N.C.: Duke University Press.
  1996a  *Culture and Environment in Inner Asia, Volume 1: The Pastoral Economy and the Environment.* Cambridge, Eng.: White Horse Press.
  1996b  *Culture and Environment in Inner Asia, Volume 2: Society and Culture.* Cambridge, Eng.: White Horse Press.

Humphrey, Caroline, Marina Mongush, and B. Telengid
  1993   Attitudes to Nature in Mongolia and Tuva: A Preliminary Report. *Nomadic Peoples* 33: 51–61.

International Decade for Natural Disaster Reduction, China National Committee (IDNDR)
  1993   *National Report of the People's Republic of China on Natural Disaster Reduction.* Available from the Editorial Department, author, no. 147 Beiheyan Street, Beijing 100721.

International Fund for Agricultural Development (IFAD)
  1989   *Completion Evaluation Report for the Northern Pasture and Livestock Development Project.* No. 0361-CH, Monitoring and Evaluation Division. Rome.
  1981   *Annual Report.* Rome.

Jackson, John Brinckerhoff
  1984    *Discovering the Vernacular Landscape.* New Haven, Conn.: Yale University Press.

Jacobson, Harold K., and Michel Oksenberg
  1990    *China's Participation in the IMF, the World Bank, and GATT: Toward a Global Economic Order.* Ann Arbor: University of Michigan Press.

Jagchid, Sechin
  1989    *Peace, War, and Trade Along the Great Wall: Nomadic-Chinese Interaction Through Two Millennia.* Bloomington: Indiana University Press.

Jagchid, Sechin, and Paul Hyer
  1979    *Mongolia's Culture and Society.* Boulder, Colo.: Westview Press.

Jankowiak, William
  1993    *Sex, Death, and Hierarchy in a Chinese City.* New York: Columbia University Press.
  1988    The Last Hurrah? Political Protest in Inner Mongolia. *Australian Journal of Chinese Affairs* 19/20: 269–288.

Jiang Chunyun
  1997    Speech before the Asian Ministerial Conference on the Implementation of the UN Convention to Combat Desertification, Beijing, May 13. Available as quoted by the Xinhua News Agency in Foreign Broadcast Information Service FBIS-TEN-97-006-L at http://wnc.fedworld.gov.

Jiang Fengqi
  1984    Shadi xiaoye jinjier guancong de shengwu liang ji qi gongyang qingkuang [Aboveground Biomass and Nutrient Contents of *Caragana microphylla* Brush on Sandy Land]. In *Fengsha ganhan zonghe zhili yanjiu: Neimenggu dongbu dichu* [Studies on the Integrated Control of Wind, Sand Drifting, and Drought in Eastern Inner Mongolia], ed. Cao Xinsun, vol. 1, pp. 100–112. Hohhot: Inner Mongolia People's Publishing House.

Jiang Hong, Zhang Peiyuan, Zheng Du, and Wang Fenghui
  1995    The Ordos Plateau of China. In *Regions At Risk: Comparisons of Threatened Environments*, ed. Jeanne X. Kasperson, Roger E. Kasperson, and B. L. Turner II, pp. 420–459. New York: United Nations University Press.

Jiang Su
  1989    Woguo caodi ziyuan heli litong yu caodi xumuye fazhan de jianyi [A Recommendation on Reasonable Utilization of Grassland Resources and Development of Animal Husbandry in China]. In *Zhongguo caodi kexue yu caoye fazhan* [Development of Grassland Science and Prataculture in China], pp. 15–18. Beijing: Science Press.

Jiang Wandi
  1994    Turning the Desert Green. *Beijing Review* 37(44): 16–20.

Jochim, Michael
    1981   *Strategies for Survival: Cultural Behavior in an Ecological Context.*
           New York: Academic Press.

Johnson, D. L.
    1979   Management Strategies for the Drylands: Available Options and
           Unanswered Questions. In *Proceedings of the Khartoum Workshop on
           Arid Lands Management*, ed. J. Mabbutt, pp. 26–35. Tokyo: United
           Nations University Press.

Jones, Francis C.
    1949   *Manchuria Since 1931.* London: Royal Institute of International Af-
           fairs.

Kernick, M. D.
    1980   *First Consultant Report on Pasture Improvement and Utilization.*
           Submitted on behalf of the Pilot Demonstration Center for Intensive
           Pasture, Fodder, and Livestock Production of Wengniute Ranch to the
           Grassland and Pasture Crops Group of the Food and Agricultural Or-
           ganization of the United Nations (FAO), Rome, May.

Khan, Almaz
    1996   Who Are the Mongols? State, Ethnicity, and the Politics of Represen-
           tation in the People's Republic of China. In *Negotiating Ethnicities in
           China and Taiwan*, ed. Melissa Brown, pp. 125–159. Berkeley: Insti-
           tute of East Asian Studies, University of California.

Khazanov, Anatoly M.
    1994   *Nomads and the Outside World.* 2d ed. Trans. Julia Crookenden.
           Madison: University of Wisconsin Press.

Knight, John, and Lina Song
    1993   The Spatial Contribution to Income Inequality in Rural China. *Cam-
           bridge Journal of Economics* 17: 195–213.

Kou Zhenwu (senior director, Wulanaodu Grassland Ecosystem Research Sta-
           tion)
    1994   Personal interview with author, Wulanaodu, July.

Kou Zhenwu and Xue Cai
    1990   Wulanaodu liusha zhili zhibei huifu [Control of Moving Sand Dunes
           and Restoration of Vegetation in Wulanaodu Region]. In *Fengsha gan-
           han zonghe zhili yanjiu: Neimenggu dongbu dichu* [Studies on the In-
           tegrated Control of Wind, Sand Drift, and Drought in Eastern Inner
           Mongolia], ed. Cao Xinsun, vol. 2, pp. 5–10. Beijing: Science Press.

Kou Zhenwu, Song Bo, Dennis Ojima, Jerrold Dodd, and Steven Williams
    1993   Model Analysis of Grassland Dynamics to Different Management
           Measures in Wulanaodu. Paper prepared for the research conference
           Grassland Ecosystem of the Mongolian Steppe, Racine, Wisconsin,
           November 4–7.

Kovda, V. A., et al.
    1979   Soil Processes in Arid Lands. In *Arid-Land Ecosystems: Structure,*

*Functioning, and Management,* ed. D. W. Goodall and R. A. Perry, vol. 1, pp. 439–470. Cambridge, Eng.: Cambridge University Press.

Lam, Willy Wo-Lap
1993  Green Movement Under Scrutiny. *South China Morning Post* (Hong Kong), July 7, p. 8.

Lao Sheh
1961  In Inner Mongolia. *Beijing Review* 46 (November 17): 12–16.

Lattimore, Owen
1994  *High Tartary.* Tokyo: Kodansha America. Original edition, Boston: Little, Brown, 1930.
1962  *Writings of Owen Lattimore, Studies in Frontier History. Collected Papers 1929–58.* New York: Oxford University Press.
1951  *Inner Asian Frontiers of China.* 2d ed. New York: American Geographical Society.
1941  *Mongol Journeys.* New York: Doubleday.
1934  *Mongols of Manchuria.* New York: The John Day Company.

Li Jianshu
1990  Preface to *A Rising Green Great Wall: Construction of the Sanbei Shelter-Forest System.* Bureau of the Ministry of Forestry in charge of construction of Shelter Forest in North, Northeast, and Northwest China, ed. Da Di Publishing House.

Li Jingli (stockyard manager, Chifeng Trading Company)
1993  Personal interview with author, October 14.

Li Peng
2000  Speech before the National People's Congress Environmental and Resources Protection Committee, Beijing, June 5. Available as quoted by the Xinhua News Agency in Foreign Broadcast Information Service FBIS-CHI-2000-0605 at http://wnc.fedworld.gov.

Li Yutang (executive vice president of the China Pratacultural Association and senior economist for China's Ministry of Agriculture)
1992  Personal interview with author, June 5.

Lin Xiangjin
1990  Development of the Pastoral Areas of Chifeng City Prefecture. In *The Wool Industry in China,* ed. John Longworth, pp. 74–92. Victoria, Australia: Inkata Press.

Little, Peter, and David Brokensha
1987  Local Institutions, Tenure, and Resource Management in East Africa. In *Conservation in Africa: People, Policies, and Practice,* ed. David Anderson and Richard Grove, pp. 193–209. Cambridge, Eng.: Cambridge University Press.

Liu Guizhen
1993  1992 nian quanguo xumuye shengchan qingkuang [Report on National Pastoral Production in 1992). *Muye tongxun* 19(5): 2–3.

Liu Yuchen
1990  Keerqin shadi de shahua guocheng ji zhonghe zhili [Sandification

Process of Keerqin Sandland and Its Comprehensive Treatment]. In *Guoji caodi zhibei xueshu huiyi lunwenji* [International Symposium on Grassland Vegetation], pp. 621–626. (Also cited in English in National Research Council 1992, pp. 60–61.)

Liu Yuman
1990   Economic Reform and the Livestock/Pasture Imbalance in Pastoral Areas of China: A Case Study of the Problems and Counter Measures in Balinyou Banner. In *The Wool Industry in China*, ed. John Longworth, pp. 93–105. Victoria, Australia: Inkata Press.

Lloyd, Evan
1986   *Hypothermia and Cold Stress*. London: Croom Helm.

Longworth, John, ed.
1990   *The Wool Industry in China*. Victoria, Australia: Inkata Press.

Longworth, John, and Gregory Williamson
1993   *China's Pastoral Region: Sheep and Wool, Minority Nationalities, Rangeland Degradation, and Sustainable Development*. Wallingford, Eng.: CAB International / Australian Centre for International Agricultural Research.

Loucks, O., M. Plumb-Menties, and D. Rogers
1985   Gap Processes and Large-Scale Disturbances in Sand Prairies. In *The Ecology of Natural Disturbance and Patch Dynamics*, ed. S. T. A. Picket and P. S. White, pp. 71–83. New York: Academic Press.

Loucks, Orie, and Wu Jianguo
1992   Xilingele. In *Grasslands and Grassland Sciences in Northern China*, ed. National Research Council, pp. 67–84. Washington, D.C.: National Academy Press.

Luo Zhewen and Zhao Luo
1986   *The Great Wall of China in History and Legend*. Beijing: Foreign Language Press.

Ma Rong
1992   Social Sciences: Chinese Literature on Grassland Studies. In *Grasslands and Grassland Sciences in Northern China*, ed. National Research Council, pp. 121–134. Washington, D.C.: National Academy Press.
1984   Migrant and Ethnic Integration in Rural Chifeng, Inner Mongolia Autonomous Region, China. Ph.D. diss., Brown University.

Mackerras, Colin
1994   *China's Minorities: Integration and Modernization in the Twentieth Century*. New York: Oxford University Press.

Mainguet, Monique
1994   *Desertification: Natural Background and Human Mismanagement*. 2d ed. New York: Springer.

Manduhu and Nasendelger
1963   Red Star of the Grasslands. *China Reconstructs* 12(2): 24–26.

Marnham, Patrick
  1979   *Nomads of the Sahel.* Minority Rights Group report no. 33. London.
McCabe, J. Terrence
  1990   Turkana Pastoralism: A Case Against the Tragedy of the Commons. In *Human Ecology* 18: 81–104.
McCay, Bonnie J., and James M. Acheson, eds.
  1987   *The Question of the Commons: The Culture and Ecology of Communal Resources.* Tucson: University of Arizona Press.
McGovern, Thomas, et al.
  1988   Northern Islands, Human Error, and Environmental Degradation: A View of Social and Ecological Change in the Medieval North Atlantic. *Human Ecology* 16(3): 225–270.
Mearns, Robin
  1993a  *Pastoral Institutions, Land Tenure, and Land Policy Reform in Post-Socialist Mongolia.* PALD research report no. 3, International Development Studies (IDS), University of Sussex, Brighton.
  1993b  Territoriality and Land Tenure Among Mongolian Pastoralists: Variation, Continuity and Change. *Nomadic Peoples* 33: 73–103.
  1991   Pastoralists, Patch Ecology, and Perestroika: Understanding Potentials for Change in Mongolia. *IDS Bulletin* 22(4): 25–33.
Meserve, Ruth
  1982   The Inhospitable Land of the Barbarian. *Journal of Asian History* 16(1): 51–89.
Meyer, Jeffrey
  1991   *The Dragons of Tiananmen: Beijing as a Sacred City.* Greenville: University of South Carolina Press.
Meyer, John, et al.
  1997   The Structuring of a World Environmental Regime, 1870–1990. *International Organization* 51(4): 623–651.
Mitchell, Timothy
  1988   *Colonising Egypt.* Cambridge, Eng.: Cambridge University Press.
Montagu, Ivor
  1956   *Land of Blue Sky.* London: Dennis Dobson.
Montell, G.
  1937   Distilling in Mongolia. *Ethnos* 2(5): 321–332.
Mortimore, Michael
  1988   Desertification and Resilience in Semi-Arid West Africa. *Geography* 73: 61–64.
Mosher, Steven
  1993   *A Mother's Ordeal.* New York: Harcourt.
Moule, Arthur
  1957   *Quinsai: With Other Notes on Marco Polo.* Cambridge, Eng.: Cambridge University Press.

Murphey, Rhoads
  1967   Man and Nature in China. *Modern Asian Studies* 1(4): 313–333.
Nan Yinhao and Wei Jun
  1990   Yinhong yuguan gailiang caochang tigao caodi hengchanli de yanjiu [Study on the Amelioration of Degraded Grassland and the Increase of Its Productivity by Warping with Diverted Floodwater]. In *Fengsha ganhan zonghe zhili yanjiu: Neimenggu dongbu dichu* [Studies on the Integrated Control of Wind, Sand Drift, and Drought in Eastern Inner Mongolia], ed. Cao Xinsun, vol. 2, pp. 66–73. Beijing: Science Press.
Nan Yinhao et al.
  1993   A Study on the Stability of the Shrub Communities as Forage and Dune Fixation. Paper prepared for the International Symposium on Grassland Resources, Hohhot, IMAR, August 16–18.
Nasihan sumu official documents
  1993   *Tongji yu shourubiao* [Year-End Township Statistics and Income Tables].
  1991   *Lizu shaqu shiji, xingjian jiating xiaocao kulun* [Construct Household Enclosures Based upon Arid-Land Realities]. Policy paper of the People's Government of Nasihan Sumu, August 15.
National Research Council, eds. (NRC)
  1992   *Grasslands and Grassland Sciences in Northern China.* Washington, D.C.: National Academy Press.
  1990   *The Improvement of Tropical and Subtropical Rangelands.* Washington, D.C.: National Academy Press.
Needham, Joseph
  1956   *Science and Civilisation in China, Volume 2: History of Scientific Thought.* Cambridge, Eng.: Cambridge University Press.
*Neimenggu ribao*
  1990   Report of the Inner Mongolia Autonomous Region Urban and Rural Construction and Environmental Protection Department. In Foreign Broadcast Information Service FBIS-CHI-90-189: 24–25.
Nelson, Ridley
  1990   *Dryland Management: The Desertification Problem.* World Bank technical paper no. 116. Washington, D.C.: World Bank.
NRC, *see* National Research Council
Nyerges, A. Endre
  1992   The Ecology of Wealth-in-People: Agriculture, Settlement, and Society on the Perpetual Frontier. *American Anthropologist* 94(4): 860–881.
Oakes, T. S.
  1993   The Cultural Space of Modernity: Ethnic Tourism and Place Identity in China. *Environment and Planning D: Society and Space* 11: 47–66.
Orleans, Leo A.
  1991   *Loss and Misuse of China's Cultivated Land: China's Economic Dilemma in the 1990's.* Submitted to the Joint Economic Committee Congress of the United States, vol. 1 (April), pp. 403–417.

Ottinger, Ulrike (film director)
   1992   *Taiga: Journey to the Northern Land of the Mongols.* 16 mm color
          film, 501 minutes. Distributed by New Yorker films, New York.
Pasternak, Burton, and Janet Salaff
   1993   *Cowboys and Cultivators: The Chinese of Inner Mongolia.* Boulder,
          Colo.: Westview Press.
Peters, Pauline
   1987   Embedded Systems and Rooted Models: The Grazing Lands of
          Botswana and the Commons. In *The Question of the Commons: The
          Culture and Ecology of Communal Resources,* ed. Bonnie J. McCay
          and James M. Acheson, pp. 171–194. Tucson: University of Arizona
          Press.
Qian Xueshen
   1984   Caoyuan, caodi he xin jishu geming [Grassland, Prataculture, and the
          Modern Technological Revolution]. In *Jishu jingji daobao* [Newspaper
          of Technological Economy], p. 1. Beijing, November 30.
Rasidondug, S., in collaboration with Veronika Veit
   1975   *Petitions of Grievances Submitted by the People (18th–beginning of
          20th century).* Wiesbaden: Otto Harrassowitz.
Reardon-Anderson, James
   1995   Man and Nature in the West Liao River Basin During the Past 10,000
          Years. Paper presented at the annual meeting of the Association for
          Asian Studies, Washington, D.C., April 6.
*Renmin ribao*
   1991   Woguo jiangda guimo zhili shamo [China Expands the Scope of De-
          sertification Control]. July 29, p. 1.
   1976   Shenru piDeng Kangzhen jiuzai [Intensify Criticism of Deng, Resist
          Quakes, and Recover from Disasters]. Aug. 11, p. 1.
Rhodes, Steven
   1991   Rethinking Desertification: What Do We Know and What Have We
          Learned? *World Development* 19(9): 1137–1143.
Rifkin, Jeremy
   1991   *Biosphere Politics.* New York: Crown.
Risser, P. G., E. C. Birney, H. D. Blocker, S. W. May, W. J. Parton, and J. A. Wiens
   1981   *The True Prairie Ecosystem.* Stroudsburg, Penn.: Hutchinson Rass.
Rokkan, S., and D. Urwin.
   1983   *Economy, Identity, Territory.* London: Sage.
Salter, Christopher, ed.
   1973   Doing Battle with Nature: Landscape Modification and Resource Uti-
          lization in the People's Republic of China, 1960–72 (annotated bibli-
          ography). Occasional paper no. 1, Asian Studies Committee, Univer-
          sity of Oregon.
Samuels, Marwyn, and Carmencita Samuels
   1989   Beijing and the Power of Place in Modern China. In *The Power of*

*Place: Bringing Together Geographical and Sociological Imaginations*, ed. John Agnew and James Duncan, pp. 202–227. Boston: Unwin Hyman.

Sandford, Stephen
1983    *Management of Pastoral Development in the Third World.* New York: John Wiley and Sons.

Schaller, George
1993. *The Last Panda.* Chicago: University of Chicago Press.

Schurmann, Herbert Franz
1956    *Economic Structure of the Yuan Dynasty* (translation of Chapters 93 and 94 of the *Yuan Shi*). Harvard–Yenching Institute Studies, vol. 16. Cambridge, Mass.: Harvard University Press.

Scoones, Ian
1996    Range Management Science and Policy: Politics, Polemics, and Pasture in Southern Africa. In *The Lie of the Land: Challenging Received Wisdom on the African Environment*, ed. Melissa Leach and Robin Mearns, pp. 34–53. Oxford: The International African Institute/James Currey.

Sears, Paul
1980    *Deserts on the March.* 4th ed. Norman: University of Oklahoma Press.

Serruys, Henry
1980    Singing Sands and Masked Dance. *Mongolian Studies: Journal of the Mongolia Society* 6: 99–102.

Severin, Tim
1991    *In Search of Genghis Khan.* London: Hutchinson.

Shen Changjiang
1985    *Pastoral Systems in Arid and Semi-Arid Zones of China.* Pastoral Network paper 13b. London: Agricultural Administration Unit, Overseas Development Institute.

Shipton, Parker
1994    Land and Culture in Tropical Africa: Soils, Symbols, and the Metaphysics of the Mundane. *Annual Review of Anthropology* 23: 347–377.

Shiva, Vandana
1995    Biotechnological Development and the Conservation of Biodiversity. In *Biopolitics: A Feminist and Ecological Reader*, ed. V. Shiva and I. Moser, pp. 193–213. London: Zed.

Simon, David
1993    The Communal Lands Question Revisited. In *Third World Planning Review* 15(1): R3–R7.

Smil, Vaclav
1993    *China's Environmental Crisis.* Armonk, N.Y.: M. E. Sharpe.
1987    Land Degradation in China: An Ancient Problem Getting Worse. In *Land Degradation and Society*, ed. Piers Blaikie and Harold Brookfield, pp. 214–222. New York: Methuen.

Smith, Richard
  1995    Getting Rich Is Glorious. *Ecologist* 25(1): 14–15.
Sneath, David
  1994    The Impact of the Cultural Revolution in China on the Mongolians of
          Inner Mongolia. *Modern Asian Studies* 28(2): 409–430.
Song Tingmin (assistant to Chen Fei, coordinator at the World Bank Project Of-
          fice of the Chinese Ministry of Domestic Trade)
  1994    Personal interview with author, Beijing, June 24.
Soong Ching Ling
  1972    How Nature Is Being Changed. *China Reconstructs* 21(4): 22–25.
*South China Morning Post* (Hong Kong)
  1994    Expert Voices Concern over Advancing Desertification. August 20, p.
          7. Also available in Foreign Broadcast Information Service FBIS-CHI-
          94-163 at http://wnc.fedworld.gov.
Stuart, Kevin
  1997    *Mongols in Western/American Consciousness.* Lewiston, N.Y.: Edwin
          Mellen Press.
Sullivan, L.
  1995    The Three Gorges Project: Dammed If They Do? *Current History*,
          September: 266–269.
Swift, Jeremy
  1996    Desertification: Narratives, Winners, and Losers. In *The Lie of the
          Land: Challenging Received Wisdom on the African Environment*, ed.
          Melissa Leach and Robin Mearns, pp. 73–90. Oxford: The Interna-
          tional African Institute / James Currey.
Szynkiewicz, Slawoj
  1982    Settlement and Community Among the Mongolian Nomads. In *East
          Asian Civilizations: New Attempts at Understanding Traditions. Vol-
          ume 1: Ethnic Identity and National Characteristics*, ed. Wolfram
          Eberhard, Krzysztof Gawlikowski, and Carl-Albrecht Seyschab, pp.
          10–44. Bremen, Germany: Simon and Magiera.
Tapp, Nicholas
  1995    Minority Nationality in China: Policy and Practice. In *Indigenous Peo-
          ples of Asia*, ed. R. H. Barnes and Benedict Kingsbury, pp. 195–220.
          Ann Arbor, Mich.: Association for Asian Studies.
Thomas, David
  1993    Sandstorm in a Teacup? Understanding Desertification. *Geographical
          Journal* 159(3): 318–331.
Thomas, David, and Nicholas Middleton
  1994    *Desertification: Exploding the Myth.* New York: John Wiley and Sons.
Timberlake, Lloyd
  1985    *Africa in Crisis: The Causes, the Cures of Environmental Bankruptcy.*
          Washington, D.C.: International Institute for Environment and Devel-
          opment.

Toynbee, Arnold
  1967    Cities in History. In *Cities of Destiny*, ed. Arnold Toynbee, pp. 13–28.
          London: Thames and Hudson.

Tsoar, Haim, and Yehuda Zohar
  1985    Desert Dune Sand and Its Potential for Modern Agricultural Develop-
          ment. In *Desert Development: Man and Technology in Sparselands*,
          ed. Yehuda Gradus, pp. 184–200. Boston: D. Reidel.

Tuan, Yi-fu
  1979    *Landscapes of Fear*. New York: Pantheon.

United Nations Environment Programme (UNEP)
  1993    *Environmental Data Report 1993–94*. Prepared for UNEP by the
          GEMS Monitoring and Assessment Research Centre, London. Lon-
          don: Blackwell Reference.
  1988    *Desertification Control Plan-Making and Implementation by the
          Lanzhou Desert Research Institute of the Academia Sinica*. Environ-
          mental Management and Planning in the PRC. Bangkok.
  1987    *Rolling Back the Desert*. Nairobi: UNEP.

Wakeman, Frederic
  1975    *The Fall of Imperial China*. New York: Free Press.

Waldron, Arthur
  1990    *The Great Wall of China: From History to Myth*. Cambridge, Eng.:
          Cambridge University Press.

Walls, James, ed.
  1982    *Combating Desertification in China*. Nairobi: United Nations Envi-
          ronment Programme.

Walter, Eugene V.
  1988    *Placeways: A Theory of the Human Environment*. Chapel Hill: Uni-
          versity of North Carolina Press.

Wang C., Wang R., Zhung S., Zhang X., and Tian L.
  1984    Wulanaodu dichu de tudi jianxing jiqi xingcheng guocheng [The Soil
          Types and Their Formation Processes in the Wulanaodu Region]. In
          *Fengsha ganhan zonghe zhili yanjiu: Neimenggu dongbu dichu* [Stud-
          ies on the Integrated Control of Wind, Sand Drifting, and Drought in
          Eastern Inner Mongolia], ed. Cao Xinsun, vol. 1. Hohhot: Inner Mon-
          golia People's Publishing House.

Wang Lixian, Wang Xian, and Zhang Kebin, eds.
  1993    *The Experiences of Combatting Desertification in P.R. China*. Beijing:
          College of Soil and Water Conservation, Beijing Forestry University.

Wang, Q. Edward
  1999    History, Space, and Ethnicity: The Chinese Worldview. *Journal of
          World History* 10(2): 285–305.

Watson, Andrew, Christopher Findlay, and Du Yintang
  1989    Who Won the "Wool War"? A Case Study of Rural Product Market-
          ing in China. *China Quarterly* 118 (June): 213–241.

WCED, *see* World Commission on Environment and Development

Wengniuteqi renmin zhengfu [People's Government of Wengniute Banner]

    1993    *Gongzuo baogao* [Work report from the 3d conference of the 11th session]. Wudan, March 24.

    1988    Wengniuteqi renmin zhengfu bangongshi wenzhang [Document no. 96 from the Office of the People's Government of Wengniute Banner]. Wudan, September 23.

    1984    Wengniuteqi renmin zhengfu wenjian [Document from the People's Government of Wengniute Banner]. Issue no. 36, article 100.

WHO, *see* World Health Organization

Williams, Dee Mack

    2000    Representations of Nature on the Mongolian Steppe: An Investigation of Scientific Knowledge Construction. *American Anthropologist* 102(3): 503–519.

    1997    Grazing the Body: Violations of Land and Limb in Inner Mongolia. *American Ethnologist* 24(4): 763–785.

Williams, Raymond

    1973    *The Country and the City.* New York: Oxford University Press.

World Bank

    1987    *China: The Livestock Sector.* Washington, D.C.: World Bank.

World Commission on Environment and Development (WCED)

    1987    *Our Common Future.* New York: Oxford University Press.

World Health Organization (WHO)

    1994    *World Health Statistics Annual, 1993.* Geneva.

Worster, Donald

    1990    The Ecology of Order and Chaos. *Environmental History Review* 14(1/2): 1–18.

Wu, David

    1990    Chinese Minority Policy and the Meaning of Minority Cultures: The example of Bai in Yunnan, China. *Human Organization* 49(1): 1–13.

Wulanaodu Gacca official document

    1993    *Hukou yu shourubiao* [Year-End Household Statistics and Income Tables].

    1984–1993    *Tongji yu xumubiao* [Statistics and Livestock Tables].

    1981    *Chengbao Tongji* [Decollectivization Statistics].

Xinhua News Agency

    2000a  PRC to Allocate $2.4 Billion for Ecological Projects in Inner Mongolia. Hohhot, IMAR, April 4. Available in Foreign Broadcast Information Service FBIS-CHI-2000-0404 at http://wnc.fedworld.gov.

    2000b  PRC to Green 4.3 Million Hectares of Sand Areas Around Beijing over 10 Years. Beijing, September 22. Available in Foreign Broadcast Information Service FBIS-CHI-2000-0922 at http://wnc.fedworld.gov.

    2000c  Zhu Rongji Calls for Immediate Curb on Expanding Deserts. Beijing,

May 22. Available in Foreign Broadcast Information Service FBIS-CHI-2000-0522 at http://wnc.fedworld.gov.

2000d  PRC's Inner Mongolia to Tackle Increased Desertification. Beijing, May 25. Available in Foreign Broadcast Information Service FBIS-CHI-2000-0525 at http://wnc.fedworld.gov.

2000e  PRC Scientists Suggest Using High-Tech to Control Desert. Beijing, April 13. Available in Foreign Broadcast Information Service FBIS-CHI-2000-0413 at http://wnc.fedworld.gov.

2000f  China Plans to Grow Improved Varieties of Grass in Space. Xian, September 20. Available in Foreign Broadcast Information Service FBIS-CHI-2000-0920 at http://wnc.fedworld.gov.

2000g  Economists Predict Narrower East-West Economic Gap. Beijing, June 15. Available in Foreign Broadcast Information Service FBIS-CHI-2000-0615 at http://wnc.fedworld.gov.

1990   Deputies Propose Forming Special Environment Committee. XNA 2601(02) N. Beijing, March 26.

1989   Responsibility System Brings Fortune to Chinese Herdsmen. Hohhot, IMAR, September 28.

1988   Foreign Funds Help Chinese Farming to Develop. Hohhot, IMAR, February 18.

1987   Change in Herdsmen's Lifestyle. Hohhot, IMAR, September 8.

1985   Herders Contract Pastures. Beijing, October 11.

Xu Bonian and Qiu Jizhou (University of Inner Mongolia)
1995   Quoted by Didi Tatlow in "Government Policy Said to Destroy Grasslands," *Eastern Express* (Hong Kong), July 2, p. 10.

Xu Lan and the Shenyang Institute of Applied Ecology
1990   Keerqin shadi xibu de jingguan tezheng [Landscape Characteristics of the Western Keerqin Sandy Lands]. *Jingguan shengtai xue* [Landscape Ecology Studies], pp. 235–238.

Xu Youfang
1997   Speech to the Asian Ministerial Conference on the Implementation of the UN Convention on Combatting Desertification, Beijing, May 15. Available as quoted by the Xinhua News Agency in Foreign Broadcast Information Service FBIS-TEN-97-135 at http://wnc.fedworld.gov.

1993   Speech to the National Desertification Conference, Chifeng, Inner Mongolia, September 24. Available as quoted by the Xinhua News Agency in Foreign Broadcast Information Service FBIS-CHI-93-184 at http://wnc.fedworld.gov.

Zev, Naveh, and Arthur Lieberman
1990   *Landscape Ecology: Theory and Application*. Student ed. New York: Springer.

Zhang Cungen
1990   An Overview of Wool Production and Processing in the Inner Mongolia Autonomous Region. In *The Wool Industry in China*, ed. John Longworth, pp. 57–73. Victoria, Australia: Inkata Press.

Zhang Xinshi
1992    Northern China. In *Grasslands and Grassland Sciences in Northern China*, ed. National Research Council, pp. 39–54. Washington, D.C.: National Academy Press.
1989    Jianli beifang caodi zhuyao leixing youhua shengtai moshide yanjiu [A Research Project on Optimized Ecological Models for Main Types of China's Northern Grasslands]. *Zhongguo guojia ziran kexue jijin zhongda xiangmu lixiangshu* [Proposal for the Important Research Project Fund, Natural Sciences Foundation of China]. Beijing.

Zhao Shidong
1992    Brief Introduction to Wulanaodu Grassland Ecosystem Research Station. Paper prepared for the research conference Grassland Ecosystem of the Mongolian Steppe, Racine, Wisconsin, March 26–29.

Zhao Songqiao
1990    The Semi-Arid Land in Eastern Inner Mongolia. *Chinese Journal of Arid Land Research* 3: 257–271.

Zhao Songqiao and Xing Jiaming
1984    Origin and Development of the Shamo and the Gobi of China. In *Evolution of the East Asian Environment, Volume 1: Geology and Palaeoclimatology*, ed. Robert Orr Whyte, pp. 230–251. Centre for Asian Studies, University of Hong Kong.

Zhao Ziyang
1981    The Present Economic Situation and the Principles for Future Economic Construction. *Beijing Review* 51: 6–32.

Zhou Li
1990    Economic Development in China's Pastoral Regions: Problems and Solutions. In *The Wool Industry in China*, ed. John Longworth, pp. 43–56. Victoria, Australia: Inkata Press.

Zhou Weide
1993    Neimeng huanjing baohu shiji [A Century of China's Environmental Protection in Inner Mongolia]. News briefing, Hohhot, October 19. Reported in *Neimenggu ribao*, October 21: 1.

Zhu Rongji
2000    Report to the Third Session of the Ninth National People's Congress, Beijing, March 5. Available as quoted by the Xinhua News Agency in Foreign Broadcast Information Service FBIS-CHI-2000-0305 at http://wnc.fedworld.gov.

Zhu Zhenda
1990    Desertification in the Northern Territory of China: Present Status and Trend of Development. *Journal of Chinese Geography* 1(1): 61–70.

Zhu Zhenda and Wang Tao
1990    Cong ruogan dianxing diqu de yanjiu dui jin shinian yu nianlai Zhongguo tudi shamohua yanbian qushi de fenxi [Analysis on the Trend of Land Desertification in Northern China During the Last

Decade Based on Examples From Some Typical Areas]. *Dili xuebao* [Acta Geographica Sinica] 45(4): 430–440.

Zimmerer, Karl
  1993   Soil Erosion and Social (Dis)courses in Cochabamba, Bolivia: Perceiving the Nature of Environmental Degradation. *Economic Geography* 69(3): 312–327.

# Index

In this index an "f" after a number indicates a separate reference on the next page, and an "ff" indicates separate references on the next two pages. A continuous discussion over two or more pages is indicated by a span of page numbers, e.g., "57–59." *Passim* is used for a cluster of references in close but not consecutive sequence.